AF762886

# HORMONES AND ENERGY METABOLISM

# ADVANCES IN EXPERIMENTAL MEDICINE AND BIOLOGY


Editorial Board:

| | |
|---|---|
| Nathan Back | *State University of New York at Buffalo* |
| N. R. Di Luzio | *Tulane University School of Medicine* |
| Bernard Halpern | *Collège de France and Institute of Immuno-Biology* |
| Ephraim Katchalski | *The Weizmann Institute of Science* |
| David Kritchevsky | *Wistar Institute* |
| Abel Lajtha | *New York State Research Institute for Neurochemistry and Drug Addiction* |
| Rodolfo Paoletti | *University of Milan* |


## Recent Volumes in this Series

# HORMONES AND ENERGY METABOLISM

Edited by
David M. Klachko
Ralph R. Anderson
and
Murray Heimberg
University of Missouri-Columbia

SPRINGER SCIENCE+BUSINESS MEDIA, LLC

Library of Congress Cataloging in Publication Data

Midwest Conference on Endocrinology and Metabolism, 12th, University of Missouri – Columbia, 1976.
Hormones and energy metabolism.

(Advances in experimental medicine and biology; v. 111)
"Proceedings of the Twelfth Midwest Conference on Endocrinology and Metabolism held at the University of Missouri, Columbia, Missouri, October 7–8, 1976, and sponsored by: University of Missouri-Columbia, College of Agriculture . . . [et al.]."
Includes index.
1. Hormones – Physiological effect – Congresses. 2. Energy metabolism – Congresses. 3. Metabolic regulation – Congresses. 3. Metabolic regulation – Congresses. I. Klachko, David M. II. Anderson, Ralph Robert, 1932- III. Heimberg, Murray. IV. University of Missouri – Columbia. College of Agriculture. V. Title. VI. Series. [DNLM: 1. Hormones – Metabolism – Congresses. 2. Energy metabolism – Congresses. Q1 AD559 v. 111/WK102.3 M629h 1976]
QP571.M52 1976 599'.01'33 78-23943
ISBN 978-1-4757-0736-6 ISBN 978-1-4757-0734-2 (eBook)
DOI 10.1007/978-1-4757-0734-2

Proceedings of the Twelfth Midwest Conference on Endocrinology and Metabolism held at the University of Missouri, Columbia, Missouri, October 7–8, 1976 and sponsored by:

University of Missouri-Columbia
- College of Agriculture
- College of Veterinary Medicine
- Dalton Research Center
- Department of Biochemistry
- Department of Medicine
- Department of Pharmacology
- Department of Physiology
- Department of Veterinary Pathology
- Division of Biological Sciences
- Extension Division
- Graduate School
- School of Medicine
- Sinclair Comparative Medical Research Farm

American Chemical Society – UMC Chapter
Ayerst Company
Brody Memorial Lectureship Committee
Ciba–Geigy Corporation
Harry S. Truman Memorial Veterans Hospital
Merck Sharp and Dohme
G. D. Searle and Company
The Upjohn Company

©1979 Springer Science+Business Media New York
Originally published by Plenum Press, New York in 1979

All rights reserved

No part of this book may be reproduced, stored in a retrieval system, or transmitted, in any form or by any means, electronic, mechanical, photocopying, microfilming, recording, or otherwise, without written permission from the Publisher

## Conference Chairmen

RALPH R. ANDERSON, Ph.D., Associate Professor of Dairy Husbandry, University of Missouri-Columbia

DAVID M. KLACHKO, M.D., Associate Professor of Medicine, University of Missouri-Columbia

## Planning Committee

JOHN D. DAVID, Ph.D., Assistant Professor of Biological Sciences, University of Missouri-Columbia

C. W. FOLEY, Ph.D., Professor of Veterinary Anatomy–Physiology, University of Missouri-Columbia

LEONARD R. FORTE, Ph.D., Associate Professor of Pharmacology, University of Missouri-Columbia

JOHN M. FRANZ, Ph.D., Associate Professor of Biochemistry, University of Missouri-Columbia

JAMES A. GREEN, Ph.D., Professor of Anatomy, University of Missouri-Columbia

LAURENCE W. HEDLUND, Ph.D., Assistant Professor of Dairy Husbandry, University of Missouri-Columbia

MURRAY HEIMBERG, M.D., Ph.D., Professor and Chairman of Pharmacology, Professor of Medicine, University of Missouri-Columbia

J. ALAN JOHNSON, Ph.D., Assistant Professor of Physiology, University of Missouri-Columbia

JOHN R. SWANSON, Ph.D., Conference Coordinator, Conferences and Short Courses, University Extension Division, University of Missouri-Columbia

WARREN L. ZAHLER, Ph.D., Assistant Professor of Biochemistry, University of Missouri-Columbia

## Speakers

JOHN H. EXTON, M.D., Ph.D., Professor of Physiology, Vanderbilt University, Nashville, Tennessee

JOHN N. FAIN, Ph.D., Professor and Chairman, Section of Physiological Chemistry, Brown University, Providence, Rhode Island

ROBERT A. HARRIS, Ph.D., Professor of Biochemistry, Indiana University School of Medicine, Indianapolis, Indiana

JOSEPH LARNER, M.D., Ph.D., Professor and Chairman, Department of Pharmacology, University of Virginia School of Medicine, Charlottesville, Virginia

ALBERT L. LEHNINGER, Ph.D., DeLamar Professor and Director, Department of Physiological Chemistry, the Johns Hopkins University School of Medicine, Baltimore, Maryland

J. DENIS MCGARRY, Ph.D., Associate Professor of Internal Medicine and Biochemistry, University of Texas Southwestern Medical School, Dallas, Texas

OLIVER E. OWEN, M.D., Professor of Medicine and Director, General Clinical Research Center, Temple University, Philadelphia, Pennsylvania

WILLIAM F. WOODSIDE, Ph.D., Assistant Professor of Pharmacology, University of New Mexico School of Medicine, Albuquerque, New Mexico

## Moderators

THOMAS W. BURNS, M.D., Professor of Medicine and Director, Division of Endocrinology, University of Missouri-Columbia

MURRAY HEIMBERG, M.D., Ph.D., Professor and Chairman of Pharmacology, Professor of Medicine, University of Missouri-Columbia

HAROLD D. JOHNSON, Ph.D., Professor and Chairman of Dairy Husbandry, University of Missouri-Columbia

BOYD L. O'DELL, Ph.D., Professor of Biochemistry, University of Missouri-Columbia

# PREFACE

The Twelfth Annual Midwest Conference on Endocrinology and Metabolism continued the tradition of selecting a topic of interest to a wide variety of scientists with interests in biology. The conference on "Hormones and Energy Metabolism" was dedicated appropriately to Dr. Samuel A. Brody, a leader in research in this field as described by Professor Johnson in this volume.

A particular feature of these conferences has been the large proportion of time devoted to discussion of each paper and the published proceedings have included edited transcripts of these discussions. Unfortunately, due to malfunction of the recording system, major portions of the discussions were lost and, despite much effort, insufficient was salvageable to be meaningful and helpful. Consequently, the editors decided to omit completely the discussions except for a communication provided by Dr. Woodside.

We again express our appreciation to the staff of Conferences and Short Courses, University of Missouri-Columbia Extension Division and especially to Dr. John Swanson who helped organize the multitude of administrative details; to Mrs. Linda Bennett, CPS, for her continuing excellent efforts in typing the manuscripts for publication; and particularly to the sponsors, without whom there would have been no conference.

Ralph R. Anderson

Murray Heimberg

David M. Klachko

CONTENTS

Dr. Samuel Brody

# DEDICATION: DR. SAMUEL A. BRODY

It is most appropriate to dedicate the Twelfth Midwest Conference on Endocrinology and Metabolism to the memory of Dr. Samuel Brody. Energy metabolism as it related to growth and development of the whole animal was the area of scientific endeavor in which Brody was so widely recognized. Intracellular energy metabolism and especially the hormonal controls have progressed very rapidly since Brody's days; however, even during the later stages of his career Brody visualized the controlling influences of hormones on the animal system to be an integrating principle of animal physiology.

Samuel Brody was professor of dairy husbandry and a member of the University of Missouri staff, 1920-1956. Words cannot describe fully the greatness of his scientific contributions. His was one of the greatest minds engaged in basic research and applied science in agriculture and the livestock industry, but his contributions covered many fields beneficial to human welfare.

He was an authority on growth, aging, metabolism, productive processes, and environmental physiology of domestic animals. He was the recipient of many honors,* and his work brought world-wide recognition to himself and to the University of Missouri.

Samuel Brody was born in Lithuania in 1890. In 1906 he emigrated to Canada, having been preceded by an older brother. He made his way as a miner, peddler, and briefly a commercial fisherman, moving then to New Hampshire where he had another older brother, I.A. Brody, still living. Several months as a machinist convinced him that he wanted to find work which would bring him into closer contact with living plants, animals, and humans. Alone and unaided he made his way to the National Farm School in Pennsyl-

---

*Guggenheim Foundation Fellowship for study in Europe, 1930-31; Recipient of the Borden Award for research in dairy production, 1950; and the National Distinguished Service Gamma Sigma Delta Award, 1955.

vania. It was here that he heard of the work of T. Brailsford Robertson, then at the University of California in Berkeley, and he set out again to study under this unique scientist.

In 1917 he received the A.B. degree with a major in biochemistry and, in 1919, the M.A. degree. In the meantime, he had joined the U.S. Air Force where he served in both the aviation and chemical warfare service until the end of the war. The following year he served as assistant biochemist in the University of California Medical School. In 1920 he was nominated by W.R. Bloor, who had succeeded Robertson on the Berkeley campus, for a position on the staff of the Department of Dairy Husbandry at the University of Missouri. The appointment was made by Professor A.C. Ragsdale, who remained as administrator and colleague to Brody for the 36 years the latter spent on the Missouri campus. As Professor of Dairy Husbandry, Brody also taught courses in the Department of Agricultural Chemistry and cooperated in the research program of that department.

In 1920 Dr. Brody married Sophie Edith Dubosky of Berkeley, who went with him to Columbia. They had two sons, Dr. Eugene B. Brody and Dr. Arnold J. Brody.

The tradition of scholarship and fundamental research was strong in the Dairy Husbandry Department at Missouri, for it had been cultivated by Dean H.J. Waters, C.H. Eckles, and L.S. Palmer. Brody found it easy and natural to take up the studies of growth and energetics which he had begun at California. His first publications on growth of dairy cows and domestic fowls appeared in the *Journal of General Physiology* as early as 1921. Practical problems on the effect of age, stage of gestation, and weight of dairy cows upon the composition and volume of milk secreted were included in these early studies.

In 1928, after study at the University of Chicago during sabbatical leave from Missouri, he completed the requirements for the Ph.D. degree under the direction of A.J. Carlson. In 1930 and 1931 he studied in five European universities on a Guggenheim Foundation Fellowship. His leaves of absence were always used for scholarly purposes.

In 1926 the Herman Frasch Foundation became interested in Brody's fundamental studies and sponsored his research in growth and development of domestic and laboratory animals for many years. Altogether 66 Agricultural Experiment Station bulletins detailing the results of the growth and development series were issued under Dr. Brody's coauthorship. In 1946 a new phase of the work was entered, the study of the influence of climatic factors on physiological reactions and productivity in farm animals. In the next

10 years Brody published 39 bulletins in this field of environmental physiology and shelter engineering.

The U.S. Department of Agriculture, the National Research Council, the U.S. Navy, and the Atomic Energy Commission were government agencies which contributed to the support of Brody's work.

In addition to his tremendous output of Experiment Station bulletins, Brody was author or coauthor of 92 journal articles which appeared between 1921 and 1956, dealing chiefly with growth, growth curves, milk secretion, energy metabolism, thyroid function, aging, evaporative cooling and effect of ambient temperature and wind on dairy cows. Some philosophical and wide ranging articles also dealt with population control and the food supply, charts for estimating profit per cow and per unit milk, and the relativity of physiologic time and weight. He contributed widely to encyclopedias and reviews, usually on growth and nutrition.

The most ambitious tour-de-force of Brody's career, however, was the compilation of his book, *Bio-Energetics and Growth, with Special Reference to the Efficiency Complex in Domestic Animals*, published in 1945 by the Reinhold Publishing Corporation of New York.

This volume consisted of 1023 pages made up of 25 chapters carrying more than 2000 references, more than 500 illustrations, and some 113 tables of data. Much of this material was taken from the Experiment Station bulletins dealing with growth and development published during the author's first 25 years in the field. Brody avowedly strove for an integrating principle "which would show by a word or a phrase the interrelatedness of all the phases of growth development and aging with the energetic efficiencies and profits in milk, meat, eggs, and muscular work production through generalizing equations," so-called for integrating unwieldy bodies of data. Such a principle might be comparable with the theory of evolution. The nearest to such an integrating principle found by Brody was the ancient, perhaps vaguely felt, concept of physiological self regulation propounded by Claude Bernard and, more recently, designated as homeostasis by W.B. Cannon. This principle, in Brody's opinion, should apply to social as well as physiological self regulation, including that of human society.

Brody defined the "living field" as the totality of the interactions in the living system with the environment, internal and external. The living field pattern, thus, resembles the electromagnetic field pattern of the physicist, and the self regulation principle of Claude Bernard resembles the theorem of LeChatelier for nonliving systems. The universality of homeostasis is illustrated in many forms, homeothermy, the maintenance of a constant level of oxygen and carbon dioxide in the blood during heavy muscular work, homeostasis of metabolism in relation to surface area, and change in form. Homeostasis of growth is established through the observations that growth proceeds as if the "normal" condition were that of mature size and that the rate of growth tends to be proportional to the distance from the mature size. Senescence is defined as the gradual failure of homeostasis, and pathology as the attempt to maintain homeostasis under conditions of injury. Reproduction may be a compensatory mechanism against senescence, maintaining constant the internal social environment in spite of the aging of its constituent members. Body weight, body water, carbohydrate level, calcium level, fat level, oxygen and acid level are all cited at length as examples of physiological homeostasis.

Social homeostasis is extended from the mores of social insects to the finer integration of human society, apparently now in a transition period due to the unbalanced development of techniques. During growth and development the living "field" is thought of as a process which organizes the diverse elements into an integrated unit. This field is stable in a dynamic rather than a static sense. The individual units, atoms, molecules, and tissues are undergoing continuous change but the pattern, "the field," remains until the homeostatic mechanism breaks down and the organism dies. This is a brief, and perhaps unfairly generalized, statement of Brody's integrating principle as set forth in his book.

A typical review of Brody's book is that of F.S. Hammett, editor of *Growth*, the periodical to which Brody contributed so much from its inception. Hammett commented as follows:

> *These thousand pages packed with charts, tables, illustrations, references to some three thousand investigators, and the scholarly analysis and comment of the author, mark the product as a classic. It is a competent, efficient, comprehensive, practical philosophical job. It is competent because it presents the data fully, concisely, logically. It is efficient because the reader can find what he wants clearly stated. It is comprehensive because it not only covers the central theme, but also the ramifications thereof. It is practical because it integrates maintenance, growth, and the products of growth with the cost thereof. And it is philosophical because it indicates how bioenergetics and growth are related to all our social, economic,*

> *political, military and associated problems. Moreover, Dr. Brody does it calmly with no drum- or breast-beating; no belaboring of any pet thesis, but just simply and convincingly. This man is a scientist.*

This biographer can only say amen to Hammett's review of the book. Seldom has so comprehensive, scholarly, and original a tome been added to our scientific heritage. No rival in its field has appeared and Brody's book will doubtless long remain the reference book par excellence on bioenergetics and growth.

Some attention is given in this book to environmental influences on energetic efficiency but most of Brody's work in this field was done after the book was published. In 1946 Brody took on a new project with strong federal support, the study of the effect of climatic conditions on the efficiency of animal energy transformations. The Psychroenergetic or Climatic Laboratory, built in 1948 on the University of Missouri campus, provided for control of temperature, humidity, air movement, light, and ventilation rate independently in the two chambers, each of which housed six mature cows or nine calves for the experimental studies. The performance of Jersey, Brown Swiss, and Holstein was compared with that of Texas-bred Brahman cows under varying conditions of temperature, light, and wind exposure. This project was a cooperative one between the U.S. Department of Agriculture and the Departments of Dairy Husbandry and Agricultural Engineering and is still in operation.

No review of the scholarly work of this man can convey the quality of his mind and personality. His colleagues at Missouri recognized his unique ability to call forth loyalty from students and also to stimulate them to strive for the best possible performance. He criticized them sternly, encouraged them to try their own ideas, but never failed to show them how to improve their product. Above all, he scorned mediocrity.

He had a remarkable sense of humor of the broad homely type characteristic of Abraham Lincoln. He never failed to enliven his lectures, whether to a class of students or to a learned society, with an unexpected and often earthy simile. There were those who misunderstood these sallies and resented them, but Brody never changed his style nor apologized for what he knew to be human and honest.

The driving force of Brody's life was his scientific ambition. Indeed, he left himself little time for anything else than his work, even during the last 20 years of life which he lived calmly and patiently under the threat of recurrent attacks of angina pectoris. He could be found in his office or laboratory seven days a week, often late at night, standing at his drawing board, writing, calculating, or drawing diagrams. The routine of laboratory measurements and techniques he left largely to his assistants and students, although he was sharply critical of both methods and results. Much of his genius lay in the ability to coordinate seemingly unrelated data and to see a logical pattern leading to mathematical development of natural relationships. His great ambition was to uncover a universal principle of biology, perhaps relating homeostasis to the principles of thermodynamics.

Based on material provided by H.D. Johnson, Chairman, Department of Dairy Husbandry and a biography by Agnes Fay Morgan, University of California, Berkeley.

# HORMONES AND
# ENERGY METABOLISM

# SOME ASPECTS OF ENERGY COUPLING BY MITOCHONDRIA

Albert L. Lehninger

Department of Physiological Chemistry
The Johns Hopkins School of Medicine
725 N. Wolfe Street
Baltimore, MD 21205

This paper will review some recent developments and trends of thought on the energetics, mechanism, and dynamics of electron transport and oxidative phosphorylation, as well as some aspects of the interplay between the mitochondrial and cytosolic compartments of cells basic to an understanding of metabolic regulation.

## Energy-linked Functions Dependent on Electron Transport

Energy derived from electron transport to oxygen is known to drive several basic mitochondrial processes in addition to oxidative phosphorylation of ADP. They include (1) reversal of electron transport in the respiratory chain, for example, from ferrocytochrome c to $NAD^+$ (2) reduction of $NADP^+$ via the energy-dependent pyridine nucleotide transhydrogenase (3) inward transport (and release) of $Ca^{2+}$ from mitochondria and (4) membrane transport between the cytosol and mitochondrial matrix of many metabolites such as phosphate, ADP, ATP, dicarboxylates, tricarboxylates, glutamate, aspartate, ornithine, pyruvate, etc., as well as $Na^+$. It is not yet possible to state the relative proportions of mitochondrial energy flow going into these secondary processes in any given cell type. At one extreme is the liver cell, in which a very complex metabolic traffic occurs between the cytosol and mitochondrial compartments (example, amino acid deamination and the urea cycle); in this case energy flow must be channeled into a number of integrated membrane transport activities, in addition to oxidative phosphorylation. At the other extreme, in the flight muscle of some insects, essentially all the respiratory energy goes into ATP production.

To return to the primary function of oxidative phosphorylation, the coupling between electron flow and ATP synthesis in intact mitochondria is energetically very efficient, since the oxidation-reduction steps of electron transport appear to be in near-equilibrium with the synthesis of extramitochondrial ATP from ADP and $P_i$. The essential reversibility of electron transport is demonstrated by the fact that ATP energy can drive electrons backwards, for example, from site II to site I, as occurs during the reduction of acetoacetate to β-hydroxybutyrate at the expense of succinate, a reaction that has been estimated to occur at rates of one-third or more of the forward rate of electron transfer and is, thus, of physiological significance, particularly in the liver (Ernster and Lee, 1964).

All the energy-requiring processes that are coupled to electron transport in mitochondria appear to utilize the same intermediate high-energy compound or state that is the immediate energy source for oxidative phosphorylation. Much evidence indicates that the different respiration-coupled processes listed above not only compete with each other for the central high-energy compound or state, but have different affinities for it. For example, when $Ca^{2+}$ is present in the surrounding medium, energy-coupled accumulation of $Ca^{2+}$ takes precedence not only over oxidative phosphorylation but all other energy-requiring processes coupled to electron flow, suggesting that $Ca^{2+}$ transport by mitochondria is a very vital process in cell activities (Lehninger, 1970). However, it appears likely that $Ca^{2+}$ transport into and out of the mitochondria in the intact cell requires only a relatively small fraction of the total energy flux, compared to ATP formation. The precise amount of electron transport energy channeled into and partitioned among the different energy-dependent mitochondrial activities presumably varies with the type of cell, with time, with location in the cell, and with local and general metabolic needs and conditions. In most cell types at least some of the total energy generated by electron transport is employed in purposes other than making ATP.

## The Energetics of ATP Synthesis in the Cell

The synthesis of three molecules of ATP from ADP and $P_i$ under standard thermodynamic conditions requires only about 3 x 7.5 = 22.5 *kcal* per mole which represents 22.5/50.8 x 100 ≅ 44% of the available standard free energy of electron transport from NADH to oxygen. However, under actual working conditions *in vitro* and presumably also in the intact cell about 16 *kcal* are required per mole of extramitochondrial ATP synthesized by mitochondria, corresponding to an efficiency of near 95% (review by Slater, 1976). The amount of energy required for ATP synthesis is assessed by determining the magnitude of the phosphorylation potential against which extramitochondrial ATP can be generated and maintained in the steady-state following an ADP-induced jump in respiratory rate. In this state four period, electron flow and ATP synthesis are in near-equilibrium. At this point the value of the phosphorylation potential ΔGp, given as

$$\Delta Gp = \Delta G^\circ + RT \ln \frac{[ATP]}{[ADP][Pi]}$$

has been found to be between 15 and 17 *kcal* in different laboratories (Slater, 1976). In short, mitochondria can keep on making ATP from ADP until very high steady-state ratios of [ATP] to [ADP] are generated. However, not all of the ∿ 16 *kcal* are used to make the ATP *per se* under these conditions. A substantial fraction of the total energy input is required to transport $ADP^{3-}$ into and $ATP^{4-}$ out of the mitochondria via the obligatorily equimolar, electrogenic $ADP^{3-}$-$ATP^{4-}$ exchange transport process (Klingenberg, 1970). The ΔGp for ATP synthesis <u>inside</u> the mitochondria is only about 12 *kcal*; in the matrix space the steady-state ratio of ATP to ADP is, in fact, lower than in the outer compartment under most conditions (Klingenberg, 1970). Therefore, some 4 *kcal* per mole are required to exchange $ADP^{3-}$ for $ATP^{4-}$ on the $ATP^{4-}$-$ADP^{3-}$ carrier of the inner membrane (Slater, 1976). The different ATP/ADP ratios are a reflection of the relatively high preference of the ATP-ADP carrier for inward transport of ADP with no preference in affinity for the two nucleotides in the outward direction (Klingenberg, 1972).

The work required to transport ATP out of the mitochondria is thus a very significant fraction, some 25%, of the total work required to make extramitochondrial ATP and, as we shall see, has not been adequately allowed for in the chemiosmotic hypothesis for oxidative phosphorylation.

It is not clear whether electron transport and oxidative phosphorylation are in near-equilibrium in the intact cell. Most investigators (*cf.* Wilson, Stubbs, Veech, Erecinska, and Krebs, 1974) favor near-equilibrium, whereas others believe that in the intact cell normally these two systems are far from equilibrium (Slater, 1976; Tager, Akerboom, Hoek, Meijer, Vaartjes, Ernster, and Williamson, 1975). The solution to this dilemma can come only from more certain knowledge of the ATP/ADP ratios in the cytosol and mitochondrial compartments of intact nonperturbed cells.

## The Electron Transport Chain

Very crucial to our understanding of the mechanism of energy transduction between oxidative and phosphorylative pathways is the identity of the electron carriers, their arrangement into a chain or chains from substrate(s) to oxygen, and their location and sidedness within the inner membrane structure. Recent research has, if anything, raised more questions than it has answered. Today a total of 16 oxidation-reduction centers appear to function in the transport of electrons from NADH to oxygen; these include seven iron-sulfur centers, a FMN flavoprotein, ubiquinone, two forms of cytochrome b($b_T$ and $b_K$), and cytochromes c, $c_1$, and a-$a_3$ (Chance, 1972; Slater, 1975). These carriers persist together in the form of complexes during purification, each complex having many polypeptide subunits. Perhaps the most significant development is the multiplicity of the iron-sulfur centers, with their characteristic oxidation-reduction potentials; their function in the chain remains obscure. The electron carriers may be arranged into four equipotential classes or pools, each having a characteristic range of midpotentials, as revealed by optically or magnetically monitored redox titrations (Chance, 1972, 1977; Wilson *et al.*, 1973). There are six equipotential carriers in pool 1, four in pool 2, four or more in pool 3, and four in pool 4. It is of interest that the pools are separated from each other by a rather large span of potential, whose position corresponds to the long-accepted sites or spans believed to furnish the energy for the three coupled phosphorylations. What is puzzling, however, is the large number of electron carriers in each of the pools. Although all respond in their oxidation-reduction state to respiratory manipulations of the mitochondria, it is not certain that all are obligatorily used during normal electron transport; parallel arrangements with cross-connections are possible. Chance has suggested that each equipotential pool functions as a redox "ballast" system, so that entry or exit of one electron does not upset the redox poise of the pool; in this view the pools represent redox buffer systems. There may be another explanation, considered below, in relationship to the charge or mass stoichiometry of coupled transport across the inner membrane.

Some years ago we proposed that there are assemblies of electron carriers in the inner membrane. This view was questioned on the grounds that the chemiosmotic hypothesis required no particular molecular ratio or close physical association of the $F_1$ ATPase with the carriers, and also on the basis that the ratio of flavoproteins to the cytochromes is nonintegral and varies in different tissues. However, the idea of a stoichiometric assembly is coming back. Slater (1975) has pointed out that there are very specific stoichiometric relationships between the $F_1$ATPase and the components of the respiratory chain between cytochromes b and $aa_3$, which strongly suggest the occurrence of stoichiometric assemblies.

The midpotentials of certain specific carriers, as observed in intact mitochondria, can change in the presence of ATP, particularly Fe·S center N-2, cytochrome $b_T$, and cytochrome $a_3$, suggesting that these carriers are most immediately implicated in energy coupling (Wilson *et al.*, 1973).

With respect to the sidedness of the electron carrier molecules, the only certainties are (1) that cytochrome c is located on the outside, (2) the $F_1$ ATPase on the inside, and (3) in the cytochrome a-$a_3$ system electrons enter from the cytosol side and protons from the matrix side; however, the path of electrons from cytochrome c to a-$a_3$ is not known. The substrate binding sites of the mitochondrial dehydrogenases and the binding site for NADH lie within the matrix compartment or on the inside aspect of the inner membrane; only the binding site of glycerol phosphate dehydrogenase is on the outside or cytosol surface.

## The $F_1$ATPase

The $F_1$ATPase, which contains some 10 subunits, is located on the inner surface anchored to a hydrophobic portion of the inner membrane via OSCP ($F_5$), the oligomycin-sensitivity conferring protine, and $F_0$, the hydrophobic protein, which appears to contain the $H^+$ channel. The latter is the site of action of oligomycin and of DCCD. Oligomycin seems to "plug" the $H^+$ channel of $F_1$ and thus inhibits the ATPase and ATP synthetase activity. The ATP is actually made on the head of the $F_1$ molecule (reviewed by Pedersen, 1975).

There are multiple binding sites for ATP and ADP on the $F_1$ molecule. Although the exact number is still uncertain, at least four sites are present, which have exceedingly high affinity for ADP and ATP. The bound nucleotides are not easily exchangeable in the nonenergized (conformational state) but are released after energization of the membrane, possibly resulting in a conformational change of the $F_1$ molecule. Slater (1974) and also Boyer (1975) have postulated that energy is required not to make the phosphate bond of ATP, but to bring about a conformational change that releases the ATP from its binding site on $F_1$. Two possible functions for the multiple binding sites have been postulated. They may be involved in a two-site, alternating "flip-flop" mechanism for ATP formation and desorption (Adolfsen and Moudrianakis, 1976) and elaborated by others (Boyer, Smith, Rosing, and Kayalar, 1975). The second idea is that one or more of the bound nucleotides is functioning as a homotropic allosteric ligand, for controlling the rate of phosphorylation.

## The Energy-coupling Mechanism

There are in principle three basic hypotheses for the mechanism of energy-transduction in oxidative phosphorylation: the chemical, chemiosmotic, and conformational coupling principles. Each ideology has some variations. Chemical coupling involves a common chemical intermediate, presumably a covalent compound with a high ΔG of hydrolysis, generated by the electron-transferring reaction and consumed in the phosphorylation reaction; the chemical coupling hypothesis at the present is not widely supported. The chemiosmotic hypothesis (Mitchell, 1966, 1968) postulates that it is the function of electron transport to generate a high-energy electrochemical gradient of protons (the protonmotive force) across the membrane, the sum of a transmembrane potential ΔΨ, negative inside, and a proton gradient ΔpH, acid outside. This gradient is postulated to be the immediate driving force for ATP synthesis. There is, however, an important alternative to the chemiosmotic hypothesis, which is deservedly gaining more attention, namely the hypothesis of R.J.P. Williams (1961), who postulated that mitochondrial energy coupling involves the movements of protons between two phases, hydrophilic and hydrophobic, within the membrane, rather than between three phases, across the membrane, as postulated by Mitchell. The conformational hypothesis (Boyer, 1965; Boyer *et al.*, 1975) assumes that the intermediate high-energy state is a special high-energy conformation of an electron carrier protein in the membrane, which is generated by oxidation or reduction and then sterically transmitted to the ATPase, where the conformational energy is used to make ATP. In this case, the energy is transmitted via the making and breaking of several or many internal noncovalent bonds of the coupling protein(s). As is indicated below, conformational changes in the $F_1$ ATPase of the mitochondrial membrane are believed to bring about energy-dependent release of ATP from the enzyme active site. It is possible that all three types of change--chemical, chemiosmotic, and conformational--take place during the total process of energy coupling, but the real crux of the matter lies in one specific part of the problem: how the energy of the oxidoreduction is converted into the energy of phosphorylation.

As things now stand, the chemiosmotic hypothesis has been and remains the most solid and useful framework for analysis of observations and ideas, particularly because it accounts comprehensively for respiration-coupled ion transport by mitochondria. However, as we shall see, it will almost certainly have to be revised in some important aspects. The balance of evidence among these hypotheses is precarious and it would require little more than a single, new, presently unpredicted insight or observation to make it all a new ball game.

## The Stoichiometry of $H^+$ Movements Coupled to Electron Transport

The stoichiometry of $H^+$ ejection coupled to electron transport can be measured under conditions in which the total electrochemical gradient generated by electron transport is converted into an equivalent $H^+$ gradient ($\Delta pH$), by collapsing the transmembrane potential with a mobile cation or anion. Usually $K^+$ is used as mobile cation together with valinomycin, which provides the $K^+$ channel through the membrane. Mitchell and Moyle (1967) concluded from their oxygen pulse measurements that two $H^+$ are ejected per pair of electrons per energy-conserving site, *i.e.*, a $H^+$/site ratio of 2.0. They also found that for each molecule of ATP hydrolyzed, 2.0 $H^+$ were ejected. From these relationships Mitchell postulated the "loop" hypothesis for the pathway of proton flow from matrix to the outside, through the alternating participation of electron-carrying and hydrogen-segments (Mitchell, 1966, 1968). Moreover, the agreement between the $H^+$/site and $H^+$/ATP ratios of 2.0 concurs with the fact that there is no net acid-base change in the ATP cycle in the cellular steady state.

However, two lines of evidence have developed which indicate that the $H^+$/site ratio of 2.0 is incorrect. One stems from the long-known fact that the $Ca^{2+}$/site ratio of electron transport is 2.0, equivalent to inward transport of four + charges per $2e^-$ per site, whereas the $H^+$/site ratio of 2.0 observed by Mitchell and Moyle involves only two positive charges moved per site (Brand, Reynafarje, and Lehninger, 1976; Rossi and Lehninger, 1964). Since $Ca^{2+}$ is now known to pass into the mitochondria by electrogenic uniport, there is a large discrepancy in the number of positive charges moved per site. The other line of evidence is thermodynamic in nature. Several investigators, beginning with Rottenberg (1970), have found that a $H^+$/site stoichiometry of 2.0 as observed by Mitchell and Moyle cannot generate a high enough total protonmotive force to make ATP at the very high phosphorylation potentials that intact mitochondria are capable of. To make a molecule of ATP at the highest phosphorylation potential requires a $\Delta Gp$ (expressed in *mV*) of at least 520 *mV*. Since the protonmotive force developed by $H^+$ ejection is 180 *mV* per $H^+$, at least 540/180 = 3.0 $H^+$ must be ejected per pair of electrons per site to generate ATP. This is, of course, a minimum value. As mentioned above, work is required to move $ADP^{3-}$ into and $ATP^{4-}$ out of mitochondria. The chemiosmotic hypothesis makes no provision for this work; it assumes that ATP-ADP exchange across the membrane is electroneutral.

This dilemma has stimulated us to re-examine the stoichiometry of $H^+$ movements coupled to electron transport and to ATP hydrolysis. In a series of papers (Brand, Chen, and Lehninger, 1976; Brand *et al.*, 1976a, 1976b; Reynafarje, Brand, and Lehninger, 1976) we have collected evidence from three different types of experimental approaches that the $H^+$/site ratio of electron transport is at least 3.0 and is, most probably, the integral value of 4.0.

*Table 1. OBSERVED STOICHIOMETRY BETWEEN $Ca^{2+}$ UPTAKE, WEAK ACID ANION ($A^-$) UPTAKE, $H^+$ EJECTION, AND OXYGEN CONSUMPTION*

| (System contained succinate and rotenone; RLM) | |
|---|---|
| $Ca^{2+}$/uptake per site | 1.90 |
| $BOH^-$/uptake per site | 4.1 |
| $BOH^-/Ca^{2+}$ ratio | 2.04 ± 0.13 |
| Net $H^+$ ejected per site | 0.1 |
| Calculated $H^+$ ejection and reuptake per site (from $A^-$ uptake) | 4.1 |

In the anion uptake method, rat liver mitochondria respiring on succinate are allowed to accumulate $Ca^{2+}$ from the medium together with a $^{14}C$-labeled weak acid anion, such as acetate, butyrate, or β-hydroxybutyrate; phosphate is excluded (Brand *et al.*, 1976). The amounts of $^{45}Ca^{2+}$ and labeled anion accumulated are measured, together with the oxygen consumption and net $H^+$ movements. Weak acid anions are used instead of phosphate because they do not pass the membrane directly but must be protonated to the free acids to enter. Since they are monocarboxylic, each anion entering carries one proton. Typical data in Table 1 show that as each pair of electrons pass each energy conserving site, two $Ca^{2+}$ ions and four $A^-$ anions are accumulated; there are no associated net $H^+$ movements. These stoichiometric findings are explained by the vectorial diagram in Figure 1, which shows that four $H^+$ must be ejected per site to account for the overall stoichiometry. All three sites seem to have the same $H^+$/site ratio of 4.0.

The second approach is through use of the oxygen pulse method of Mitchell and Moyle (1967) with some refinements. In this procedure a known amount of oxygen is injected into an anaerobic suspension of mitochondria exposed to substrate. The ensuing ejection of $H^+$ is estimated, following correction for the rapid exponential decay of the pH gradient generated by the burst of electron flow. We were easily able to reproduce Mitchell and Moyle's data and obtained a H+/site ratio of 2.0 under their conditions. However, we found that this method seriously underestimates the H+/site ratio because of an interfering secondary ion transport process which causes very rapid back-movement of $H^+$ into the mitochondria together with phosphate from the medium, via the $H_2PO_4^-$-$OH^-$ exchange or its formal equivalent, the inward co-transport of $H_2PO_4^-$ and $H^+$ (Brand *et al.*, 1976a, b). During anaerobic preincubation prior to the oxygen pulse, phosphate leaks from the mitochondria. When the

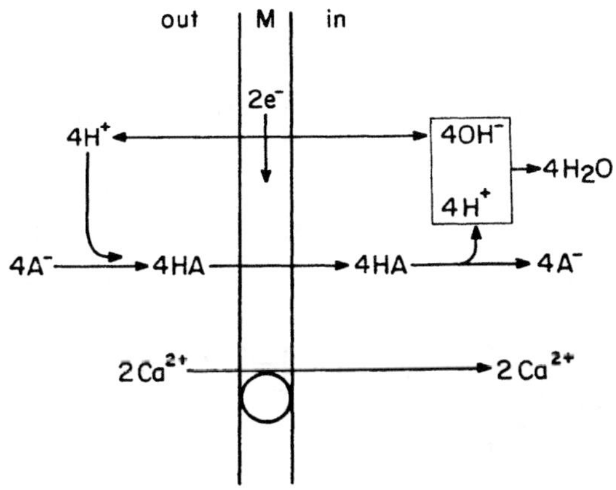


*Figure 1.* *Vectorial diagram for uptake of $Ca^{2+}$ and weak acid anion ($A^-$). The stoichiometric data in Table 1 can only be accounted for as shown, i.e., by a $H^+$/site ratio of 4.0, since $Ca^{2+}$ is now known to be transported inward by divalent electrogenic uniport.*

oxygen pulse is added, the net effect is that one $H^+$ is pulled back into the mitochondria with each $H_2PO_4^-$. When the transport of phosphate is eliminated, either by inhibition of the phosphate carrier with N-ethylmaleimide, by prior depletion of phosphate from the mitochondria, or by running the reaction at 5°, at which the phosphate movements are greatly slowed, then the $H^+$/site ratio measured by the oxygen pulse method turns out to be at least 3.0. Because the oxygen-pulse method has some other undesirable features which may also be a cause of underestimation, we believe the true $H^+$/site ratio is higher than 3.0.

The third method we have devised (Reynafarje *et al.*, 1976) involves measurement of the steady-state initial rates of $H^+$ ejection and oxygen consumption following addition of a substrate to deenergized but aerobic mitochondria. From the slopes of the electrode traces we again get a value of 2.0 for the $H^+$/site ratio, but after removal of the interfering phosphate movements, the $H^+$/site ratio was observed to be between 3.5 and 4.0. This result has been obtained with two-site and three-site substrates and with mitochondria from several species of cells. From these recent experiments we have concluded that at each of the three energy-conserving sites at least 3.0 and more likely 4.0 $H^+$ are ejected per pair of electrons, or twice as many as Mitchell has assumed. We have also reevaluated the $H^+$/ATP ratio by an improved ATP-pulse method and have found its value to be at least 2.0, possibly as high as 3.0 (Brand and Lehninger, 1977).

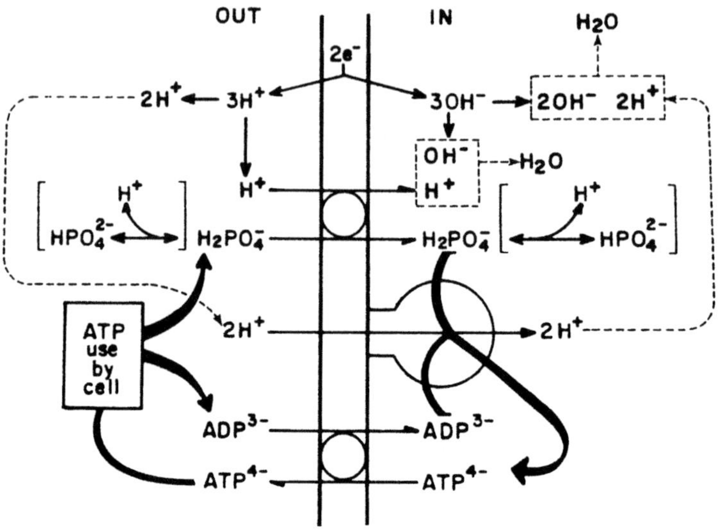


*Figure 2. Vectorial diagram of a modification of the chemiosmotic hypothesis. In this scheme, which incorporates recent stoichiometric data indicating that the $H^+$/site ratio is greater than 2.0, it is assumed that the $H^+$/site ratio is 3.0. Note that two $H^+$ are used to generate the high energy phosphate bond of intramitochondrial ATP and that one $H^+$ is used for the transport work required to move ATP outward in exchange for ADP and $P_i$. If the $H^+$/site ratio is actually 4, as is believed, then three $H^+$ would be used to make the ATP and one $H^+$ to transport it.*

These new data now allow a more satisfactory accounting (Brand and Lehninger, 1977) for the thermodynamics and stoichiometry of oxidative phosphorylation (Figure 2). We propose that at least three $H^+$ are ejected per site. One of the ejected $H^+$ is used to bring about the transport of ADP into and ATP out of the mitochondria, by providing the means to counterbalance the electrogenic nature of this process via the $H_2PO_4^-$-$OH^-$ exchange. The other two (or three) $H^+$ are utilized to drive the synthesis of the ~P bond of the new ATP molecule formed in the matrix. Thus, about 1/4 to 1/3 of the total protonmotive force would be used to drive the transport of ATP; this agrees with the thermodynamic analysis, which predicts that about 1/4 of the total $\Delta G_p$ is used to drive ATP transport (see above).

A second implication of the newer stoichiometric data, which have also been supported by other laboratories (Wiechmann, Beem, and van Dam, 1975), is that the loop hypothesis proposed by Mitchell for the mechanism of the $H^+$ ejection process will have to be radically modified or abandoned, since it can yield only two $H^+$ per $2e^-$ per site. A probable alternative is that the four $H^+$ transported per site arise from a membrane Bohr effect, in which four Bohr protons are taken up and lost vectorially across the membrane as a pair of electrons moves into and out of the energy-coupled electron carrier molecule, accompanied by conformational changes.

## Further Analysis of the Mechanism of Oxidative Phosphorylation

The chemiosmotic hypothesis, although modified significantly as just described, still represents the most useful framework for analysis and comparison of data and observations. However, we must recognize that strong as the circumstantial evidence for it may be, it may have to be modified further, and possibly radically, to accommodate other newly emerging data.

One very basic question is whether the protonmotive force generated by electron transport is, indeed, the immediate driving force for oxidative phosphorylation. It is possible (Figure 3) that it is not, but is in labile equilibrium with some high-energy chemical or conformational intermediate. We know with certainty that electron flow or ATP hydrolysis can generate protonmotive force and that protonmotive force can generate ATP and reverse the direction of electron flow, but we have as yet no proof that these transformations are direct. If they are indirect, then some type of conformational coupling might well be involved, in which Bohr protons are released or absorbed directionally across the membrane during conformational changes in the electron carriers accompanying energization and deenergization of the carriers, as indicated above.

Another basic question is which of the two energy terms in the electrochemical potential is primary, *i.e.*, is generated first, the transmembrane electric potential or the $H^+$ gradient. Here the available evidence indicates that in photosynthetic electron trans-

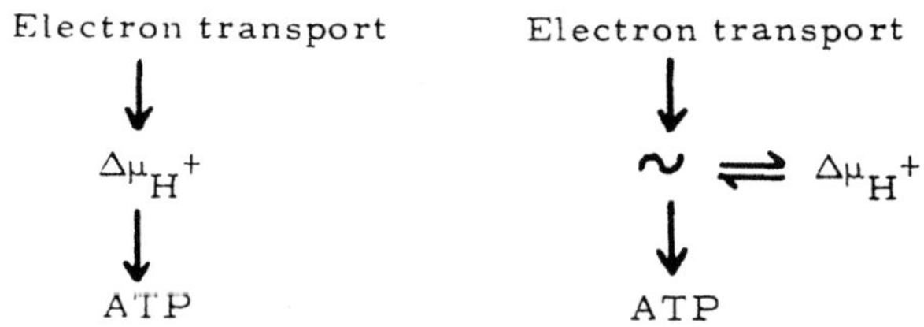


*Figure 3. The mitochondriac's dilemma.*

port and in the cytochrome a-$a_3$ system of mitochondria the transmembrane potential appears first, the $H^+$ movements being much slower. Separation of charge from mass, thus, may be the basic pattern; $H^+$ movements can be used as stoichiometric measures only under conditions in which the membrane potential is collapsed and converted into $H^+$ gradients.

A third and most basic consideration, now the subject of intense debate and thought, is whether the electrochemical gradient of $H^+$ developed by electron transport occurs across the membrane, *i.e.*, across three phases, as postulated by Mitchell, or between two phases, *i.e.*, between hydrophilic and hydrophobic zones within the membrane, as proposed by R.J.P. Williams, who has postulated that the appearance of $H^+$ in the extra-mitochondrial and $OH^-$ in the matrix compartments during electron flow are side reactions which dump $H^+$ and $OH^-$ into "sinks" not normally used in oxidative phosphorylation (Williams, 1975).

## Regulation of Electron Transport and ATP Formation

Although it is now widely appreciated that the rate of synthesis of ATP by mitochondria is determined primarily by the metabolic need for ATP, the actual molecular mechanism of respiratory regulation is still obscure. In nearly every well-studied metabolic pathway, as I have tried to point out in a textbook (Lehninger, 1975), we find that biological regulation of metabolism is made possible by the occurrence of two oppositely-directed enzymatic pathways, each of which is essentially irreversible and each of which is under separate allosteric control, so that the synthetic and degradative paths can be independently regulated, as in the case of glycogen synthesis and degradation. Allosteric regulation is, thus, a reflection of kinetic control of pathways that are unidirectional, *i.e.*, essentially irreversible and thus far removed from thermodynamic equilibrium.

However, the mechanism of regulation of electron transport and ATP synthesis does not appear to conform to this well-known pattern. We now know that electron flow from substrates to oxygen is freely reversible at all levels except possibly the span from cytochrome $a_3$ to oxygen; moreover, as indicated above, the oxidative phosphorylation of ADP is in near-equilibrium with electron transport. Reversal of electron flow almost certainly occurs in intact cells. So far as we know, however, the pathway of carriers taken by electrons flowing toward oxygen is the same as the path taken by electrons when they are pushed back to the reducing end of the respiratory chain by a high phosphorylation potential. Thus electron transport and coupled phosphorylation do not seem to occur by opposing participation of unidirectional pathways, independently regulated through kinetic control. Thus, we cannot expect that the electron transport-phosphorylation system is kinetically, *i.e.*, allosterically, controlled since inhibition of any single step in the reversible chain would result in equal degrees of inhibition for both the forward and reverse reactions.

Most of the recent thought about the regulation of respiration and phosphorylation has been based on a different type of regulation; mass-action or thermodynamic regulation, rather than the usual kinetic regulation seen in other metabolic pathways. Thus, whenever the concentration of ADP rises, or better, the phosphorylation potential declines, electron flow toward oxygen increases, with concomitant increase in ATP synthesis, until the phosphorylation potential comes into equilibrium with the oxidation-reduction state of the respiratory chain. Conversely, whenever the phosphorylation potential is suddenly increased, electrons tend to flow backward. This is the current picture of the mechanism of regulation of respiration and ATP synthesis considered strictly at the level of the electron transport chain (*cf.* Tager, Akerboom, Hoek, Meijer, Vaartjes, Ernster, and Williamson, 1975; Wilson *et al.*, 1974).

It is not certain that the regulation of respiration and ATP synthesis by mass-action effects really makes good engineering sense, compared to kinetic regulation. It appears more likely that there are still some aspects of electron transport and oxidative phosphorylation that we do not understand, aspects that have been ignored because of the preoccupation of most investigators with the mechanism of energy coupling and the structure of the inner membrane. It will be very important, in my opinion, to look more closely into the components and arrangement of the electron transport chain for evidence of at least one step that is irreversible or that is promoted by two independent, unidirectional but opposing steps that are capable of independent control. Such a step in the electron transport chain might well be regulated, in one direction or the other, through kinetic, *i.e.*, allosteric effectors, which might include not only homotropic action of ADP, ATP, or $P_i$, but heterotropic effects of other key metabolites, such as phosphoenolpyruvate. In any case, the regulation of respiration and phosphorylation is a problem that is central to all aspects of metabolic regulation considered in this Symposium.

## ACKNOWLEDGEMENTS

This work was supported by grants from the National Institutes of Health (GM05919) and the National Science Foundation (BMS 75-21923) and a Contract (NO1-CP-45610) from the National Cancer Institute.

## REFERENCES

Adolfsen, R. and Moudrianakis, E. (1976). Binding of adenine nucleotides to the purified 13S coupling factor of bacterial oxidative phosphorylation. *Arch. Biochem. Biophys.* 172:425-433.

Boyer, P.D. (1965). Carboxyl activation as a possible common reaction in substrate-level and oxidative phosphorylation and in muscle contraction. In: T.E. King, H.S. Mason, and M. Morrison (Eds.) *Oxidases and Related Redox Systems*. pp. 994-1008. Wiley, New York.

Boyer, P.D., Smith, D.J., Rosing, J., and Kayalar, C. (1975). Bound nucleotides and conformational changes in oxidative and photophosphorylation. In: E. Quagliariello, S. Papa, F. Palmieri, E.C. Slater, and N. Siliprandi (Eds.) *Electron Transfer Chains and Oxidative Phosphorylation*. pp. 361-372. North-Holland, Amsterdam.

Brand, M.D. and Lehninger, A.L. (1977). H+/ATP ratio during ATP hydrolysis by mitochondria: Modification of the chemiosmotic theory. *Proc. Nat. Acad. Sci.* 74:1955-1959.

Brand, M.D., Chen, C-H, and Lehninger, A.L. (1976). Stoichiometry of H+ ejection during respiration-dependent accumulation of $Ca^{2+}$ by rat liver mitochondria. *J. Biol. Chem.* 251:968-974.

Brand, M.D., Reynafarje, B., and Lehninger, A.L. (1976a). Reevaluation of the H+/site ratio of mitochondrial electron transport with the oxygen pulse technique. *J. Biol. Chem.* 251:5670-5679.

Brand, M.D., Reynafarje, B., and Lehninger, A.L. (1976b). Stoichiometric relationship between energy-dependent proton ejection and electron transport in mitochondria. *Proc. Nat. Acad. Sci.* 73:437-441.

Chance, B. (1977). Electron transfer: Pathways, mechanisms, and controls. *Ann. Rev. Biochem.* 46:967-980.

Chance, B. (1972). The nature of electron transfer and energy coupling reactions. *FEBS Lett.* 23:3-20.

Ernster, L., and Lee, C.P. (1964). Biological oxidoreductions. *Ann. Rev. Biochem.* 33:729-788.

Klingenberg, M. (1972). ATP synthesis and adenine nucleotide transport in mitochondria. In: S.G. Van den Bergh, P. Borst, L.L.M. Van Deenen, J.C. Riemersma, E.C. Slater, and J.M. Tager (Eds.) *Mitochondria: Biogenesis and Bioenergetics. Biomembranes: Molecular Arrangements and Transport Mechanisms*. pp. 147-162. North-Holland, Amsterdam.

Klingenberg, M. (1970). Metabolite transport in mitochondria: An example for intracellular membrane function. *Essays in Biochem.* 6:119-159.

Lehninger, A.L. (1975). *Biochemistry*. Worth Publishers, Inc., New York.

Lehninger, A.L. (1970). Mitochondria and calcium ion transport. *Biochem. J.* 119:129-138.

Mitchell, P. (1968). *Chemiosmotic Coupling and Energy Transduction.* Glynn Research Ltd., Bodmin, England.
Mitchell, P. (1966). Chemiosmotic coupling in oxidative and photosynthetic phosphorylation. *Biol. Rev.* 41:445-502.
Mitchell, P. and Moyle, J. (1968). Proton translocation coupled to ATP hydrolysis in rat liver mitochondria. *Eur. J. Biochem.* 4:530-539.
Mitchell, P. and Moyle, J. (1967). Respiration-driven proton dislocation in rat liver mitochondria. *Biochem. J.* 105: 1147-1162.
Pedersen, P.L. (1975). Mitochondrial adenosine triphosphatase. *J. Bioenergetics* 6:243-275.
Reynafarje, B., Brand, M.D., and Lehninger, A.L. (1976). Evaluation of the $H^+$/site ratio of mitochondrial electron transport from rate measurements. *J. Biol. Chem.* 251:7442-7451.
Rossi, C.S. and Lehninger, A.L. (1964). Stoichiometry of respiratory stimulation, accumulation of $Ca^{++}$ and phosphate, and oxidative phosphorylation in rat liver mitochondria. *J. Biol. Chem.* 239:3971-3980.
Rottenberg, H. (1970). ATP synthesis and electrical membrane potential in mitochondria. *Eur. J. Biochem.* 15:22-28.
Slater, E.C. (1976). Intra- and extramitochondrial phosphorylation potentials. In: J.M. Tager, H.D. Soling, and J.R. Williamson (Eds.), *Use of Isolated Liver Cells and Kidney Tubules in Metabolic Studies*. pp. 65-77. North-Holland, Amsterdam.
Slater, E.C. (1975). Components of electron-transfer chains. In: E. Quagliariello *et al.* (Eds.) *Electron Transfer Chains and Oxidative Phosphorylation.* pp. 3-14. North-Holland, Amsterdam.
Slater, E.C. (1974). Electron transfer and energy conservation. In: L. Ernster, R.W. Estabrook, and E.C. Slater (Eds.) *Dynamics of Energy-Transducing Membranes*. pp. 1-20. Elsevier, Amsterdam.
Tager, J.M., Akerboom, T.P.M., Hoek, J.B., Meijer, A.J., Vaartjes, W., Ernster, L., and Williamson, J.R. (1975). Ammonia and energy metabolism in isolated mitochondria and intact liver cells. In: F.A. Hommes and C.J. Van den Berg (Eds.) *Development of Energy Metabolism.* pp. 63-72. Academic Press, New York.
Wiechmann, A.H.C.H., Beem, E.P., and van Dam, K. (1975). The relationship between $H^+$ translocation and ATP synthesis in mitochondria. In: E. Quagliariello *et al.* (Eds.) *Electron Transfer Chains and Oxidative Phosphorylation.* pp. 335-342. North-Holland, Amsterdam.
Williams, R.J.P. (1975). Protein connections between protons, electrons, and ATP. In: E. Quagliariello *et al.* (Eds.) *Electron Transfer Chains and Oxidative Phosphorylation.* pp. 417-422. North-Holland, Amsterdam.

Williams, R.J.P. (1970). Electron transfer, conformation changes and energy conservation. In: J.M. Tager, S. Papa, E. Quagliariello, and E.C. Slater (Eds.) *Electron Transport and Energy Conservation*. pp. 7-23. Adriatica, Bari.

Williams, R.J.P. (1961). Possible functions of chains of catalysts. *J. Theor. Biol.* 1:1-17.

Wilson, D.F., Stubbs, M., Veech, R.L., Erecinska, M., and Krebs, H.A. (1974). Equilibrium relations between the oxidation-reduction reactions and the adenosine triphosphate synthesis in suspensions of isolated liver cells. *Biochem. J.* 140:57-64.

# HORMONAL CONTROL OF HEPATIC LIPOGENESIS[1]

R.A. Harris, J.P. Mapes[2], R.S. Ochs, D.W. Crabb, and L. Stropes

Department of Biochemistry
Indiana University School of Medicine
Indianapolis, IN 46202

OUTLINE



---

[1]Original research of authors supported by grants from the American Heart Association (Indiana Affiliate), Showalter Residuary Trust, and NIH AM 19259.

[2]Present Address: Metabolic Research Laboratory, Nuffield Department of Clinical Medicine, Radcliffe Infirmary, Oxford, OX2 6HE, England.

2. mechanism of glucagon and $Bt_2cAMP$ inhibition of aerobic glycolysis
3. relationship between the activity of the pyruvate dehydrogenase system and fatty acid synthesis
4. inhibitory site(s) of cyclic AMP beyond aerobic glycolysis

IV. Summary

## INTRODUCTION

Hepatic lipogenesis is subject to both fine (short-term) and coarse (long-term) control mechanisms. Coarse control, *i.e.*, adaptive change in enzyme quantity, is recognized as being important for nutritional and hormonal control of lipogenesis (for reviews see Gibson, Lyons, Scott, and Muto, 1972; Goodridge, 1975; Lakshmanan, Nepokroeff, and Porter, 1972; Numa and Yamashita, 1974). Hepatic enzymes of lipogenesis which appear to respond to this mode of regulation include acetyl-CoA carboxylase (E.C. 6.4. 1.2), fatty acid synthetase system, stearyl CoA desaturase system, glucokinase (E.C. 2.7.1.2), glucose-6-phosphate dehydrogenase (E. C. 1.1.1.49), pyruvate kinase (E.C. 2.7.1.40), phosphofructokinase (E.C. 2.7.1.11), and the citrate cleavage enzyme (E.C. 4.1.3.8), and malic enzyme (E.C. 1.1.1.38).

Fine control, *i.e.*, modulation of enzymatic activity by substrate supply, by allosteric effectors, and by interconversion between active and inactive forms by phosphorylation-dephosphorylation, is less well understood as a means for regulation of hepatic lipogenesis, but is clearly established as important because the rate of lipogenesis fluctuates over a much greater range than the adaptive change in levels of lipogenic enzymes (Veech and Guynn, 1974). Attention will be focused here on the fine control of lipogenesis. Glucagon inhibits fatty acid synthesis by liver slices (Meikle, Klain, and Hannon, 1973; Tepperman and Tepperman, 1972), by isolated hepatocytes (Geelen and Gibson, 1975, 1976; Goodridge, 1973a; Watkins and Lane, 1976), by perfused liver (this report, but also see Raskin, McGarry, and Foster, 1974), and by liver *in vivo* (Nielsen, Mehlman, Veech, and Hawkins, 1974). Since $Bt_2cAMP$[3] has similar effects in all of these liver preparations, this action of glucagon is presumably mediated through generation of endogenous cyclic AMP[4]. This effect of cyclic AMP has been the subject of several investigations (Allred and Roehrig, 1972, 1973; Capuzzi, Rothman, and Margolis, 1974; Geelen and Gibson, 1975, 1976;

---

[3]$Bt_2cAMP$: $N^6,O^{2'}$-Dibutyryl Adenosine-3':5'-monophosphate.

[4]Cyclic AMP: Adenosine-3':5'-monophosphate.

Goodridge, 1973a; Harris, 1975; Harris and Yount, 1975; Watkins and Lane, 1976). A number of enzymatic steps have been implicated but our understanding of the mechanism involved remains incomplete. Indeed, since glucagon, insulin, and presumably cyclic AMP appear to be involved in adaptive changes in the quantity of the lipogenic enzymes (Gibson *et al.*, 1972), concern has been expressed that short-term inhibition of lipogenesis by cyclic AMP may represent an artifact of *in vitro* experiments. One artifact of such studies has been described (Harris and Yount, 1975). The incubation of liver slices with cyclic AMP was found to increase the concentration of intracellular adenine nucleotides (AMP, ADP, and ATP). This expansion of the adenine nucleotide pool size contributes to the inhibition of lipogenesis and complicates mechanistic studies; however, $Bt_2$cAMP and glucagon inhibit lipogenesis without expanding the adenine nucleotide pool size (Harris and Yount, 1975).

## ENZYMATIC SITES OF FINE CONTROL OF HEPATIC FATTY ACID SYNTHESIS

Potential sites of inhibition of fatty acid synthesis by cyclic AMP are briefly discussed below. For a more comprehensive discussion, the reader is referred to reviews by Goodridge, 1975; Katz and Rognstad, 1976; Lane, Moss, and Polakis, 1974; Numa and Yamashita, 1974; and Veech and Guynn, 1974. The putative pathway for fatty acid synthesis by the liver is in Figure 1.

### Glycolysis

Excessive carbohydrate intake has long been known to promote hepatic lipogenesis (Zakim, Pardini, Herman, and Sauberlich, 1967). Furthermore, the glycolytic capacity of the liver from well-fed rats is sufficient to support its normal capacity for *de novo* fatty acid synthesis (Brunengraber, Boutry, and Lowenstein, 1973; Woods and Krebs, 1971). It follows that much of the pyruvate (and acetyl CoA) used for hepatic lipogenesis is generated within the liver by glycolysis either from exogenous glucose or endogenous glycogen. Nevertheless, exogenous lactate and pyruvate have been shown to increase greatly the rate of fatty acid synthesis by various liver preparations (Clark, Rognstad, and Katz, 1974; Goodridge, 1973b; Harris, 1975; Salmon, Bowen, and Hems, 1974). This means that the capacity of the liver to generate lactate and pyruvate can, indeed, limit hepatic lipogenesis and that factors or conditions which inhibit aerobic glycolysis should also inhibit *de novo* fatty acid synthesis in the absence of another source of acetyl CoA.

Glucagon and $Bt_2$cAMP inhibit aerobic glycolysis, *i.e.*, the accumulation of lactate and pyruvate from endogenous glycogen and exogenous glucose. This has been observed with liver slices (Meikle *et al.*, 1973; Metz and Salter, 1962), isolated hepatocytes (Harris, 1975; Seglen, 1973), and the perfused liver (Glinsmann, Hern, Linarelli, and Farese, 1969). This effect is due to an in-

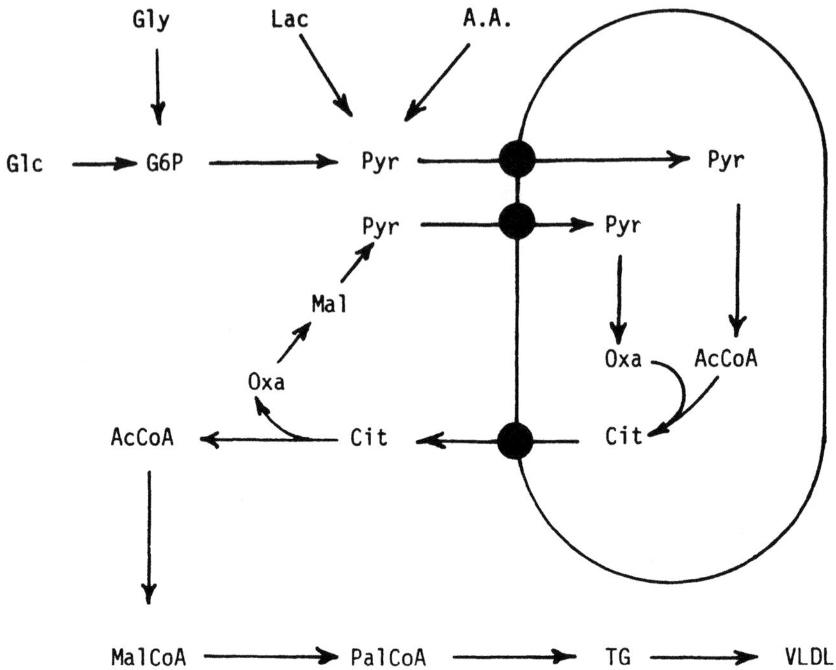


*Figure 1. Pathway believed involved in de novo fatty acid synthesis by the liver. The reducing equivalents contributed by isocitrate dehydrogenase and the pentose phosphate pathway are not indicated. Key: Gly=glycogen; Glc=glucose; G6P=glucose-6-phosphate; Lac=lactate; A.A.=amino acids; Pyr=pyruvate; AcCoA=acetyl-CoA; Oxa=oxaloacetate; Cit=citrate; Mal=malate; Mal CoA=malonyl-CoA; Pal CoA=palmityl-CoA; TG=triglycerides; VLDL=very low density lipoproteins.*

hibition of net utilization of glucose for aerobic glycolysis (Harris, 1975) as well as the activation of lactate gluconeogenesis (Glinsmann *et al.*, 1969). Although neither the sites nor the mechanism of inhibition are firmly established, much evidence suggests regulation at the level of phosphofructokinase and pyruvate kinase. Besides being regulated by multiple effectors, these enzymes are subject to regulation by a phosphorylation-dephosphorylation mechanism. A direct link between interconversion and the cyclic AMP system has been established for pyruvate kinase but not for phosphofructokinase. Ljungström, Hjelmquist, and Engström (1974) demonstrated that rat hepatic pyruvate kinase is inactivated by phosphorylation and activated by dephosphorylation. In contrast, Brand and Söling (1975) demonstrated rat hepatic phosphofructokinase is activated by phosphorylation and inactivated by dephosphorylation. The opposite response to phosphorylation of these enzymes is difficult to reconcile with an inhibition of both enzymes by a cyclic AMP-dependent mechanism. Nevertheless, in the experiments of Taunton, Stifel, Green, and Herman (1974), glucagon injected directly into the portal vein rapidly produced conversion of both enzymes into less active forms. Glucagon and cyclic AMP have also been found to diminish the activity of pyruvate kinase with perfused liver (Blair, Cimbala, Foster, and Morgan, 1976) and isolated hepatocytes (Feliu, Hue, and Hers, 1976; van Berkel, Kruijt, Koster, and Hulsmann, 1976). Likewise, in studies with isolated hepatocytes, Veneziali and Swenson (1975) found inactivation of phosphofructokinase by monobutyryl cyclic AMP.

## Pyruvate Dehydrogenase System

The pyruvate dehydrogenase system is subject to regulation by interconversion, being inactive in the phosphorylated mode and active in the dephosphorylated mode (Linn, Pettit, and Reed, 1969; Patzelt, Löffler, and Weiland, 1973). In addition, the complex is subject to product inhibition by acetyl CoA (Garland and Randle, 1964). Interconversion of the complex plays an important role in hormonal regulation of lipogenesis in adipose tissue (Berman and Halperin, 1973; Taylor and Jungas, 1974); however, the importance of interconversion in the regulation of hepatic tissue lipogenesis is less well established (Crabb, Mapes, Boersma, and Harris, 1976; Seiss and Wieland, 1976). Cyclic AMP does not appear to be directly involved in regulating either the kinase or the phosphatase which carry out the interconversion of the pyruvate dehydrogenase system (Mapes and Harris, 1976). However, cyclic AMP may indirectly influence the degree of phosphorylation by causing changes in the intramitochondrial ATP/ADP ratio, the NAD/NADH ratio, the free $Ca^{++}$ concentration, and the pyruvate concentration and, hence, the activity of the pyruvate dehydrogenase system (Batenburg and Olson, 1975; Pettit, Pelley, and Reed, 1975; Severson, Denton, Bridges, and Randle, 1976; Siess and Wieland, 1975, 1976; Walajtys, Gottesman, and Williamson, 1974). As an example of this type of regulation, cyclic AMP might increase hepatic long-chain acyl-CoA levels (discussed below), which might inactivate the pyruvate dehydrogenase system indirectly *via* inhibition of the adenine nucleotide transport system and elevation of the intramitochondrial ATP/ADP ratio (Löffler, Bard, and Wieland, 1975).

## Citrate Transport

The transport of citrate from the mitochondrion is catalyzed by an antiport system which is subject to inhibition by long-chain acyl-CoA (Halperin, Cheema-Dhadli, Taylor, and Fritz, 1974). Inhibition of the egress of citrate from the mitochondrion would diminish the generation of cytoplasmic acetyl CoA and decrease the allosteric activating effect of citrate on acetyl-CoA carboxylase. In support of this hypothesis, Halperin *et al.* (1974) found the Km of the mitochondrial antiport for citrate to be a function of the nutritional or hormonal state of the animal. Mitochondria isolated from the liver of fed rats had lower levels of long-chain acyl-CoA and a lower Km for citrate transport than mitochondria isolated from fasted or diabetic rats. This may explain, in part, the decreased capacity of hepatic mitochondria from fasted, diabetic, or fat-fed rats to support lipogenesis in a cell-free, reconstituted system (Mapes and Harris, 1975). On the other hand, Veech and Guynn (1974) calculated that the cytoplasmic acetyl-CoA/CoA ratio is always high and that there should not be a shortage of cytoplasmic acetyl-CoA under a variety of nutritional conditions. Further insight into this possible site of regulation may only be gained by direct determination of cytoplasmic citrate and acetyl-CoA levels.

## Acetyl-CoA Carboxylase

Allred and Roehrig (1973) demonstrated an inactivation of acetyl-CoA carboxylase of liver slices incubated with $Bt_2cAMP$ or cyclic AMP. Inactivation of this enzyme in the liver was also reported to be caused by the administration of glucagon *in vivo* (Klain and Weiser, 1973). Likewise, Lee, Thrall, and Kim (1973) found that $Bt_2cAMP$ caused the inactivation of acetyl-CoA carboxylase of epididymal fat tissue. The explanation for these results may have been provided by the studies of Carlson and Kim (1974), which demonstrated that acetyl-CoA carboxylase is an interconvertible enzyme, being active in the dephosphorylated mode and inactive in the phosphorylated mode. However, no evidence has yet been presented that the kinase responsible for the phosphorylation and inactivation of acetyl-CoA carboxylase is sensitive to cyclic AMP. Also, Lane *et al.* (1974) and Watkins and Lane (1976) reported that they were unable to demonstrate cyclic AMP inactivation of acetyl-CoA carboxylase with intact liver preparations. Furthermore, Halestrap and Denton (1974) failed to find any evidence for a cyclic AMP-dependent, phosphorylation-mediated inactivation of the acetyl-CoA carboxylase of adipose tissue. Obviously, additional studies are needed to clarify a possible conversion of acetyl-CoA carboxylase of the liver into an inactive form by a cyclic AMP-dependent mechanism.

Citrate is a positive allosteric effector whereas long-chain acyl-CoA is a negative effector of acetyl-CoA carboxylase (see review of Lane *et al.*, 1974); thus, acetyl-CoA carboxylase may be subject to feed-forward activation by citrate and feed-back inactivation by long-chain acyl-CoA. In short-term refeeding experiments of fasted rats Nishikori, Iritani, and Numa (1973) demonstrated an immediate increase in hepatic fatty acid synthesis which correlated with an increase in citrate and a decrease in long-chain acyl-CoA. In studies with cells isolated from chicken liver, Goodridge (1973b) also found the rate of fatty acid synthesis to correlate directly with the level of citrate and inversely with small changes in the level of long-chain acyl-CoA. These studies support the view that these effectors are responsible for short-term regulation of fatty acid synthesis *via* their opposing effects upon acetyl-CoA carboxylase. Brunengraber *et al.* (1973) found a positive correlation between citrate and the rate of fatty acid synthesis by the perfused liver. Watkins and Lane (1976) reported studies with isolated chicken hepatocytes in which both glucagon and $Bt_2cAMP$ caused a greater than 90% reduction in the level of citrate and a corresponding inhibition of fatty acid synthesis. However, in studies with rats maintained under nutritional conditions that established a wide range of rates of hepatic lipogenesis, Guynn, Veloso, and Veech (1972) found only the level of long-chain acyl-CoA and not the level of citrate to correlate with the rate of lipogenesis. Moreover, the rapid inhibition of fatty acid synthesis by glucagon *in vivo* was not found to correlate with a decrease in citrate nor with an increase in long-chain acyl-CoA (Nielsen *et al.*, 1974). Likewise, in studies from this laboratory (Harris, 1975), $Bt_2cAMP$ inhibition of fatty acid synthesis by isolated rat hepatocytes was not observed to correlate with a decrease in the level of citrate.

The concentration of citrate required for the activation of acetyl-CoA carboxylase is considerably greater (Ka = 3-4 mM) than the apparent physiological concentration of citrate (0.2-0.5 mM) (Lane *et al.*, 1974). It has been argued that conditions within the cell modify the concentration of citrate required for activation and also that citrate may be compartmentalized between the cytoplasm and mitochondria. A procedure has recently been developed for directly measuring the cytoplasmic and mitochondrial concentrations of the metabolites of isolated hepatocytes (Zuurendonk and Tager, 1974). Application of this procedure to the problem of the relative importance of citrate in the regulation of acetyl-CoA carboxylase should yield informative results.

## Fatty Acid Synthetase System

Fatty acid synthetase has been reported by Qureshi, Jenik, Kim, Lornitzo, and Porter (1975) to undergo a phosphorylation cycle, being inactive in the phosphorylated mode and active in the dephosphorylated mode. However, any link with cyclic AMP remains to be established; in fact, Klain and Weiser (1973) did not observe a decrease in the activity of the enzyme with animals injected with glucagon.

Guynn *et al.* (1972) demonstrated that the *in vivo* concentration of malonyl-CoA is low with respect to the Km of the fatty acid synthetase complex, and that, *in vivo* a positive correlation exists between rates of fatty acid synthesis and levels of malonyl-CoA. Thus, the rate of fatty acid synthesis appears determined by the level of malonyl-CoA, which sets the velocity of the reaction catalyzed by the fatty acid synthetase. Presumably, the level of malonyl-CoA is determined by the activity of acetyl-CoA carboxylase, which in turn may be determined by substrate supply, effector control, or interconversion by the phosphorylation-dephosphorylation mechanism.

## Fatty Acid Esterification and Oxidation

Free fatty acids inhibit *de novo* fatty acid synthesis by liver and other tissues (Mays and Topping, 1974; McGee and Spector, 1975; Nilsson, Sundler, and Akesson, 1973). This feedback inhibition prevents acetyl-CoA units generated by β-oxidation from being reutilized for fatty acid synthesis. The inhibitory effect of free fatty acids may be explained, as discussed in detail above, by long-chain acyl-CoA inhibition of acetyl-CoA carboxylase, citrate transport, or the pyruvate dehydrogenase system. Thus, lipogenesis is dependent upon the liver's capacity to dispose of free fatty acids and long-chain acyl-CoA. Presumably any factor (such as cyclic AMP) which influences the utilization of long-chain acyl-CoA has the potential to modify hepatic lipogenesis.

The liver of the well-fed animal uses long-chain acyl-CoA primarily for triglyceride synthesis, whereas the liver of the fasted animal consumes these esters by oxidation and ketogenesis. This switch in utilization may be explained, in part, by reciprocal changes in the activities of glycerophosphate acyltransferase (E.C. 2.3.1.15) and carnitine acyltransferase (E.C. 2.3.1.21) (see the reviews of Bremer, Bjerve, Christophersen, Daae, Solberg, and Aas, 1974; van Tol, 1975). The carnitine acyltransferase on the inner surface of the mitochondrial inner membrane appears the most important regulatory site (McGarry and Foster, 1974). Furthermore, the hepatic concentration of carnitine determines its capacity for ketogenesis. Carnitine has been shown to stimulate ketogenesis by liver slices (Fritz, 1955), isolated hepatocytes (Christiansen, Berrebaek, and Bremer, 1976), and perfused liver (McGarry, Robles-Valdes, and Foster, 1975a). A direct correlation between the rate of ketogenesis and the carnitine content of the perfused liver was observed by McGarry *et al.* (1975a). How carnitine levels of the liver are modulated remains to be established.

Glucagon and/or $Bt_2cAMP$ stimulate ketogenesis by the perfused liver of the fed rat in some studies (Heimberg, Weinstein, and Kohout, 1969; Weinstein, Klausner, and Heimberg, 1973) but not in others (McGarry, Wright, and Foster, 1975b). Triglyceride secretion by the liver appears to be inhibited concurrently with the shunting of fatty acids into ketone body synthesis (Heimberg *et al.*, 1969). The mechanism responsible for this action of cyclic AMP has not been established. It may be an effect upon the enzymes involved in the partitioning between esterification and β-oxidation or it might result from cyclic AMP mobilization of liver glycogen. A negative correlation is known to exist between the liver's capacity for ketogenesis and the hepatic content of glycogen (McGarry *et al.*, 1975b).

The liver may also contain a lipase responsive to cyclic AMP (Claycomb, Bynagle, and Kilsheimer, 1969). Activation of such a lipase could prevent net triglyceride formation by "futile cycling" triglyceride, free fatty acids, and long-chain acyl-CoA. This might also establish a higher steady state concentration of long-chain acyl-CoA and lead to inhibition of acetyl-CoA carboxylase, citrate transport, and the pyruvate dehydrogenase system. Additional studies on this potential mechanism of regulation of hepatic lipogenesis are necessary.

## RECENT EXPERIMENTAL OBSERVATIONS ON CYCLIC AMP INHIBITION OF LIPOGENESIS

The experiments presented below were carried out for the most part by methods and procedures described in detail previously (Crabb *et al.*, 1976; Harris, 1975; Mapes and Harris, 1976). Hepatocytes were isolated from the livers of rats meal-fed three hours daily on a high carbohydrate diet or from livers of 7- to 10-day-old chickens. The only important recent modification of the procedure was artificial ventilation of the anesthetized rat or chicken during cannulation of the liver by means of a Harvard rodent respirator using 98% oxygen-2% $CO_2$. Cells were prepared from chicken liver by the same procedure used for rat liver with the exception that back perfusion by way of the inferior vena cava was used because of difficulty encountered in cannulating the portal vein. Rates of fatty acid synthesis were estimated by the incorporation of $^3H_2O$ into the fatty acids of the total lipid fraction of the cells.

### Relationship between Glycolysis and Fatty Acid Synthesis

Table 1 illustrates the effect of $Bt_2cAMP$ on the metabolic properties of isolated hepatocytes prepared from meal-fed rats. The expected increase of glycogen utilization and in glucose release from the cells due to glycogen phosphorylase activation was observed. Moreover, net glucose utilization (*i.e.*, glycogen utilized corrected for glucose release into the medium) was markedly inhibited by $Bt_2cAMP$ (Harris, 1975). This inhibition of net glucose utilization is a major factor in preventing the accumulation of lactate (Table 1) and pyruvate (not shown). Thus, $Bt_2cAMP$ inhibits aerobic glycolysis by rat liver cells. Fatty acid synthe-

*Table 1. EFFECT OF $Bt_2cAMP$ ON GLUCOSE UTILIZATION AND FATTY ACID SYNTHESIS BY HEPATOCYTES ISOLATED FROM MEAL-FED RATS*

| Rate of | $-Bt_2cAMP$ | $+Bt_2cAMP$ |
|---|---|---|
| | μmol/min/g | |
| Glycogen utilization | 2.7 ± 0.1 | 3.8 ± 0.3[a] |
| Glucose accumulation | 1.2 ± 0.1 | 3.7 ± 0.3[a] |
| Net glucose utilization | 1.5 ± 0.2 | 0.1 ± 0.1[a] |
| Lactate accumulation | 1.5 ± 0.1 | 0.02 ± 0.08[a] |
| Fatty acid synthesis | 0.33 ± 0.02 | 0.07 ± 0.01[a] |

*Rates are given for the time period between 15 and 30 min after initiation of the incubation of the hepatocyte suspensions. Tritiated water incorporation into the fatty acid of the total lipid extract was used as the estimate of the rate of fatty acid synthesis. $Bt_2cAMP$ was added at an initial concentration of 50 μM.*

*The results are expressed as mean ± standard error of the mean with three liver cell preparations in each group. Data from Harris (1975).*

[a]*$P < 0.05$ by Student's t-test for paired data.*

sis, as measured by $^3H_2O$ incorporation, is inhibited concurrently with the suppression of aerobic glycolysis by $Bt_2cAMP$ (Table 1).

Glucagon has effects similar to $Bt_2cAMP$ on the metabolism of isolated hepatocytes. As shown in Figure 2, glucagon effectively inhibits net glucose utilization, lactate and pyruvate accumulation, and fatty acid synthesis. All of these effects closely parallel one another as a function of glucagon concentration, with the half maximally effective concentration being about $10^{-8}M$ and the threshold at the more 'physiological' concentration of $10^{-10}$ to $10^{-9}M$.

The perfused liver was found to respond qualitatively to glucagon in nearly the same manner as isolated hepatocytes (Table 2); however, some tendency was noted for less sensitivity of the perfused liver to glucagon compared to isolated hepatocytes. Although the explanation for this difference in sensitivity is not apparent at this time, it could involve: (a) damage to the isolated hepatocytes during preparation; (b) the difference in the glycogen content of the two preparations (considerable glycogen is lost during the isolation of the hepatocytes); and (c) the greater ratio of

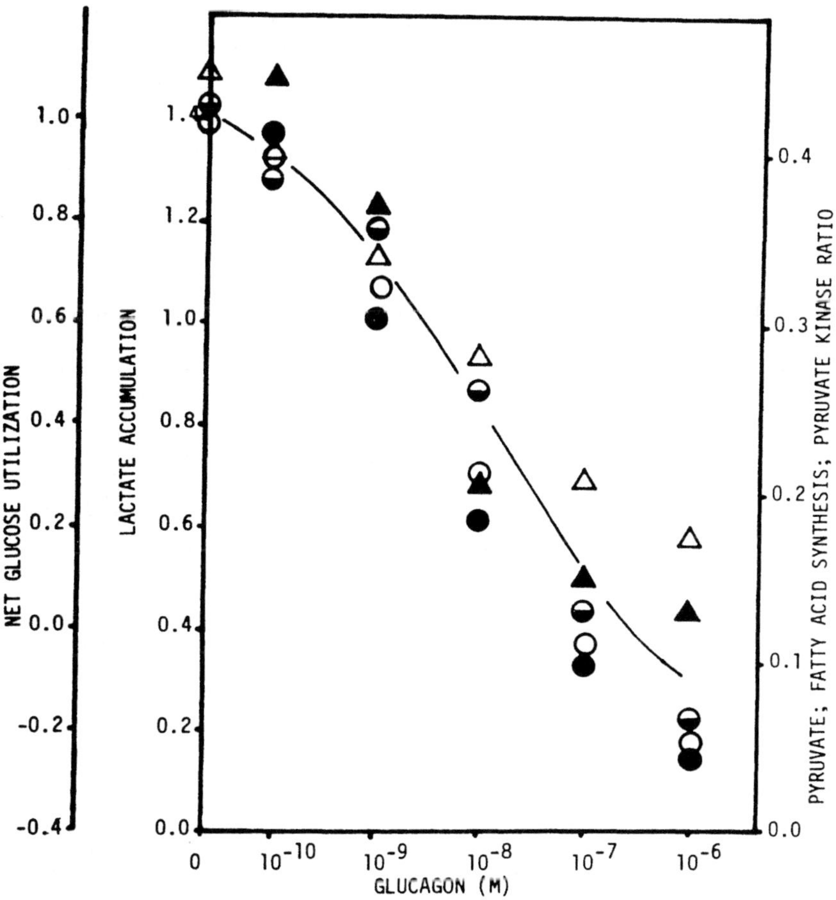


*Figure 2. Effect of glucagon on net glucose utilization, i.e., glycogen used minus glucose released, µmol/min/g (O); lactate accumulation, mM (●); pyruvate accumulation, mM (◒); fatty acid synthesis by $^3H_2O$ incorporation, µmol/min/g (Δ); and pyruvate kinase activity as low to high ratio as described in Table 3 (▲). Unpublished results of Harris and Ochs.*

medium volume to cell concentration used in the isolated hepatocyte incubations. The latter may be an important factor because, as described below, exogenous lactate and pyruvate partially overcome the inhibition of fatty acid synthesis by $Bt_2cAMP$ and glucagon. Previous studies from this laboratory have shown that the magnitude of lactate plus pyruvate accumulation is a function of the quantity of cells incubated and that this influences the rate of lipogenesis observed (Harris, 1975).

*Table 2. EFFECT OF GLUCAGON ON GLYCOGENOLYSIS, AEROBIC GLYCOLYSIS, AND FATTY ACID SYNTHESIS BY THE PERFUSED RAT LIVER*

| Measurement | No addition | Glucagon |
|---|---|---|
| Glycogen (*μmol/g* wet wt) | 229 ± 16 | 87 ± 23[a] |
| Lactate (*mM*) | 2.6 ± 0.3 | 1.2 ± 0.3[b] |
| Pyruvate (*mM*) | 0.36 ± 0.04 | 0.11 ± 0.03[a] |
| Fatty acid synthesis (*μmol* "$C_2$"/*min/g* wet wt) | 0.34 ± 0.07 | 0.16 ± 0.02[b] |

*Tritiated water incorporation into the fatty acids of the total lipid was used to estimate the rate of fatty acid synthesis. It was added to the perfusion medium at zero time. Glucagon was added initially at a concentration of* $10^{-6}$ M *and then infused continuously at a rate of 100 nmoles/hr. Analysis was performed on samples obtained after 1 hr of perfusion.*

*The results are expressed as mean ± standard error of the mean with the livers from eight animals in each group. Unpublished observations of Harris and Mapes.*

[a]*P < 0.001 by Student's t-test.*

[b]*P < 0.02 by Student's t-test.*

$Bt_2cAMP$ and glucagon also inhibit lipogenesis and net glucose utilization by isolated hepatocytes prepared from 7- to 10-day-old chickens (results not shown); however, a major difference has been noted between chicken and rat hepatocytes. Chicken hepatocytes accumulate very little lactate or pyruvate relative to that accumulated by rat hepatocytes. The rate of glycolysis seems more nearly equal to the rate of fatty acid synthesis in the case of the chicken hepatocytes. This is a striking difference because in our experience rat hepatocyte preparations which do not accumulate large amounts of lactate and pyruvate (*e.g.*, because of depleted glycogen stores) always exhibit low rates of lipogenesis. However, chicken hepatocytes have the capacity to carry out lipogenesis at rapid rates at lactate and pyruvate concentrations which would not support fatty acid synthesis by isolated hepatocytes from the rat. This interesting difference between chicken and rat hepatocytes is being investigated.

## Mechanism of Glucagon and $Bt_2cAMP$ Inhibition of Aerobic Glycolysis

Cross-over analysis has been applied to the problem of cyclic AMP inhibition of aerobic glycolysis (Harris, 1975). The metabolite

*Table 3. EFFECT OF GLUCAGON AND $Bt_2cAMP$ ON THE ACTIVITY OF PYRUVATE KINASE OF HEPATOCYTES*

| Addition | Pyruvate kinase Low PEP | Pyruvate kinase High PEP | Low/high ratio |
|---|---|---|---|
| None | 23 ± 3 | 63 ± 5 | 0.36 ± 0.01 |
| Glucagon | 8.1 ± 0.8[a] | 59 ± 4 | 0.13 ± 0.01[a] |
| $Bt_2cAMP$ | 9.6 ± 1.5[a] | 63 ± 5 | 0.14 ± 0.02[a] |

*Pyruvate kinase activity was determined 15 min after initiation of the incubation. $Bt_2cAMP$ was added at an initial concentration of 0.05 mM; glucagon at $10^{-6}$ M. An aliquot (10 µl) of the cell suspension was added to a pyruvate kinase assay mixture (1.0 ml) which contained 136 µmoles of Tris Cl (pH 7.5), 66 µmoles of KCl, 10 µmoles of $MgSo_4$, 2.5 µmoles of Mg ADP, 0.16 µmoles of NADH, 3.2 units of lactate dehydrogenase, and 10 mg of Triton X-100. The reaction was initiated by the addition of either 1.3 µmoles of phosphoenolpyruvate (low PEP rate) or 6.6 µmoles of phosphoenolpyruvate (high PEP rate).*

*The results are expressed as mean ± standard error of the mean with three rat liver cell preparations in each group. Results of Ochs and Harris (1977).*

[a]*$p < 0.05$ by Student's t-test for paired data.*

profile shows $Bt_2cAMP$ to increase fructose-6-phosphate and phosphoenolpyruvate and to decrease fructose-1,6-diphosphate and pyruvate, *i.e.*, cross-overs are apparent at the levels of phosphofructokinase and pyruvate kinase. This type of analysis only indicates that the apparent effectiveness of these enzymes is decreased by cyclic AMP. The actual explanation may be either a diminished activity of these enzymes or an increase in the activity of the opposing glucoenogenic enzymes, *i.e.*, fructose diphosphatase, phosphoenolpyruvate carboxykinase, and pyruvate kinase can be explained, in part, by a decrease in fructose-1,6-diphosphate (Harris, 1975), a positive effector of the enzyme. In addition, Table 3 shows that incubation of isolated hepatocytes from meal-fed rats with either Bt cAMP or glucagon causes a decrease in the activity of pyruvate kinase when assayed at subsaturating concentrations of phosphoenolpyruvate. This effect on the apparent $K_m$ of the enzyme for phosphoenolpyruvate is similar to that reported by others (see section II-1) and probably explains, in large part, the cross-over in the second half of the glycolytic pathway. Further evidence that inactivation of pyruvate kinase by cyclic AMP is an important factor in limiting the rate of aerobic glycolysis is presented in Figure 2. Here is shown that the extent of inactivation of pyruvate kinase caused by varying concentrations of glucagon closely parallels the extent of inhibition of net glucose utilization, lactate accumulation, and fatty acid synthesis.

The difference noted above between rat and chicken liver cells caused us to investigate the regulation of pyruvate kinase in chicken liver cells. No evidence was found for a cyclic AMP-dependent alteration of the $K_m$ of the chick liver enzyme for phospho*enol*pyruvate (Ochs and Harris, 1977). Rat liver has the L isozyme of pyruvate kinase, whereas chicken liver has the K isozyme (Strandholm, Cardenas, and Dyson, 1975). It seems likely that the more primitive K isozyme is not interconverted to a less active form by a cyclic AMP-dependent mechanism.

Phosphofructokinase is an interconvertible enzyme which may also respond to cyclic AMP (see Section II-1). However, all attemps in this laboratory to show any cyclic AMP-dependent change in activity of this enzyme of either chicken or rat hepatocytes have been unrewarded. This includes assay of the enzyme under conditions of near saturating or subsaturating substrate concentrations. Perhaps regulation of flux at this level by cyclic AMP occurs indirectly *via* effectors and/or through changes in activity of fructose diphosphatase.

Relationship between the Activity of the Pyruvate Dehydrogenase System and Fatty Acid Synthesis

The pyruvate dehydrogenase system is inactivated by phosphorylation and activated by dephosphorylation (see Section II-2). In Table 4 a comparison is made between the activity of the dephosphorylated form of the enzyme (active form) with the measured rate of fatty acid synthesis. In general, there is a good correlation between the two under a variety of incubation conditions. Furthermore, conditions which increase the pyruvate concentration of the incubation medium result in activation of the pyruvate dehydrogenase system and greater rates of fatty acid synthesis. This includes the incubation of hepatocytes with glucose in high concentration, fructose, dihydroxyacetone, and lactate plus pyruvate. Likewise, incubation conditions which decrease the pyruvate concentration of the incubation medium result in an inactivation of the pyruvate dehydrogenase system and correspondingly lower rates of fatty acid synthesis. This includes incubation of the cells with $Bt_2cAMP$, oleate, and glycerol. These results can be explained by the known inhibition of the pyruvate dehydrogenase kinase by pyruvate. Thus, the pyruvate concentration of the medium determines the activity of the pyruvate dehydrogenase system which, in turn, plays an important role in providing acetyl-CoA for fatty acid synthesis. The only condition in which there is low pyruvate and high rates of fatty acid synthesis is incubation with dichloroacetate, which itself inhibits the kinase and activates the pyruvate dehydrogenase system. Hence, the correlation between high pyruvate dehydrogenase activity and high rates of fatty acid synthesis still

*Table 4. PYRUVATE DEHYDROGENASE, PYRUVATE CONCENTRATION, AND THE RATE OF FATTY ACID SYNTHESIS*

| Addition | Fatty acid synthesis | Pyruvate dehydrogenase | Pyruvate concentration |
|---|---|---|---|
| | μmol "$C_2$"/min/g | | mM |
| None | 0.36 ± 0.07 | 0.65 ± 0.05 | 0.31 ± 0.02 |
| Glucose | 0.70 ± 0.08[a] | 0.82 ± 0.10[a] | 0.41 ± 0.07[a] |
| Fructose | 0.74 ± 0.08[a] | 1.09 ± 0.07[a] | 0.44 ± 0.05[a] |
| Dihydroxyacetone | 0.88 ± 0.06[a] | 0.99 ± 0.07[a] | 0.36 ± 0.06[a] |
| Lactate + Pyruvate | 0.84 ± 0.07[a] | 1.1 ± 0.1[a] | 0.78 ± 0.07[a] |
| $Bt_2$cAMP | 0.05 ± 0.01[a] | 0.37 ± 0.07[a] | 0.02 ± 0.01[a] |
| Oleate | 0.17 ± 0.03[a] | 0.24 ± 0.07[a] | 0.06 ± 0.08[a] |
| Glycerol | 0.11 ± 0.04[a] | 0.36 ± 0.08[a] | 0.03 ± 0.01[a] |
| Dichloroacetate | 0.43 ± 0.05[a] | 1.4 ± 0.1[a] | 0.10 ± 0.04[a] |

*Tritiated water was added to the incubation medium after 30 min of preincubation. Incubations were terminated after 45 min for the determination of pyruvate and pyruvate dehydrogenase activity; after 60 min for the determination of the rate of fatty acid synthesis. The concentrations of the additions were: $Bt_2$cAMP (0.05 mM), lactate (10 mM), pyruvate (1 mM), oleate (2 mM), glucose (45 mM), fructose (8 mM), dihydroxyacetone (10 mM), glycerol (10 mM), and dichloroacetate (1 mM).*

*The results are expressed as mean ± standard error of the mean for at least four liver cell preparations. Unpublished results of Mapes and Harris.*

[a]*$P < 0.05$ by the Student's t-test for paired samples.*

holds true.

By inhibiting aerobic glycolysis, $Bt_2$cAMP prevents the accumulation of pyruvate and indirectly causes the inactivation of the pyruvate dehydrogenase system. This appears to be an important event in the overall mechanism responsible for cyclic AMP inhibition of lipogenesis.

## Inhibitory Site(s) of Cyclic AMP beyond Aerobic Glycolysis

Substrate supplementation experiments have been used in an attempt to further localize the sites of cyclic AMP inhibition of lipogenesis. This approach is restricted, however, to substrates such as lactate, pyruvate, and acetate which readily penetrate the plasma membrane of the hepatocyte. As shown in Table 5, exogenous lactate plus pyruvate stimulates lipogenesis by rat liver cells and overcomes, to a considerable extent, the inhibition caused by $Bt_2$cAMP. This is consistent with a major site of $Bt_2$cAMP action at the level of aerobic glycolysis. On the other hand, exogenous

*Table 5. EFFECT OF LACTATE PLUS PYRUVATE AND ACETATE ON $Bt_2cAMP$ INHIBITION OF FATTY ACID SYNTHESIS AS MEASURED BY $^3H_2O$ AND $^{14}C$-ACETATE INCORPORATION*

| Addition | Rate of fatty acid synthesis -$Bt_2cAMP$ | +$Bt_2cAMP$ | Percent inhibition |
|---|---|---|---|
| | *µmol/min/g* | | |
| None | 0.27 ± 0.07 | 0.06 ± 0.02[a] | 76 |
| Acetate | 0.33 ± 0.06 | 0.06 ± 0.02[a] | 83 |
| | (0.15 ± 0.01) | (0.02 ± 0.01[a]) | 87 |
| Lactate + Pyruvate | 0.67 ± 0.08 | 0.46 ± 0.06[a] | 31 |
| Acetate + Lactate + Pyruvate | 0.68 ± 0.07 | 0.51 ± 0.06[a] | 25 |
| | (0.31 ± 0.02) | (0.26 ± 0.02[a]) | 16 |

*Rates of fatty acid synthesis were determined between 30 and 60 min of incubation by $^3H_2O$ and $^{14}C$-acetate incorporation. Results obtained for $^{14}C$-acetate incorporation are given in parentheses. The initial concentrations of the added components were: lactate, 10 mM; pyruvate, 1 mM; $^{14}C$-acetate added at 30 min, 10 mM; and $Bt_2cAMP$, 50 µM.*

*The results are expressed as mean ± standard error of the mean with three rat liver cell preparations in each group. Data similar to that published by Harris (1975).*

[a]*$p < 0.05$ by Student's t-test for paired data.*

acetate fails to overcome the inhibition of fatty acid synthesis caused by $Bt_2cAMP$. This result is obtained whether fatty acid synthesis is measured by $^3H_2O$ or $^{14}C$-acetate incorporation (Table 5). Since acetate is readily activated in the cytoplasm of rat liver (Barth, Sladek, and Decker, 1971), no shortage of cytoplasmic acetyl-CoA for fatty acid synthesis should occur in the presence of exogenous acetate. Thus, supplementation experiments with acetate do not support the conclusion that $Bt_2cAMP$ inhibition of lipogenesis is simply a consequence of the inhibition of aerobic glycolysis. However, lactate plus pyruvate partially reverses $Bt_2cAMP$ inhibition of the incorporation of $^{14}C$-acetate into fatty acids (Table 5). Thus, lactate and/or pyruvate is required for $^{14}C$-acetate incorporation into fatty acids, and inhibition of aerobic glycolysis by $Bt_2cAMP$ does explain at least in part the inhibition of fatty acid synthesis as measured by either $^3H_2O$ or $^{14}C$-acetate incorporation.

The requirement for lactate and/or pyruvate for hepatic fatty acid synthesis has been further investigated wtih other inhibitors of lipogenesis (Tables 6 and 7). A somewhat analogous approach to the problem was used by Goodridge (1973b). α-Cyano-4-hydroxycin-

*Table 6. EFFECT OF CYANOHYDROXYCINNAMIC ACID (CHC), 5-TETRADECYLOXY-2-FUROIC ACID (TOFA) AND (-)HYDROXYCITRATE (OHC) ON FATTY ACID SYNTHESIS*

| Added substrate | Added inhibitor | Rate of fatty acid synthesis *μmol/min/g* | Percent inhibition |
|---|---|---|---|
| None | None | 0.39 ± 0.07 | -- |
| | CHC | 0.07 ± 0.02[a] | 83 |
| | TOFA | 0.06 ± 0.01[a] | 85 |
| | OHC | 0.145 ± 0.005[a] | 63 |
| Lactate + Pyruvate | None | 0.99 ± 0.08 | -- |
| | CHC | 0.14 ± 0.04[a] | 85 |
| | TOFA | 0.08 ± 0.01[a] | 92 |
| | OHC | 0.24 ± 0.03[a] | 75 |

*Rates of fatty acid synthesis were determined between 30 and 60 min of incubation by $^3H_2O$ incorporation. The initial concentrations of the added components were: lactate, 10 mM; pyruvate, 1 mM; α-cyano-4-hydroxycinnamic acid, 2 mM; 5-tetradecyloxy-2-furoic acid, 0.05 mM; and (-) hydroxycitrate, 2 mM.*

*The results are expressed as mean ± standard error of the mean with three or four rat liver cell preparations in each group. Unpublished results of Harris.*

[a]*$p < 0.05$ by Student's t-test.*

namic acid inhibits pyruvate uptake by the mitochondrion (Halestrap, 1975); 5-tetradecyloxy-2-furoic acid inhibits citrate transport from the mitochondrion (Kariya, Parker, Grisar, Martin, and Wille, 1975; Panek, Cook, and Cornell, 1976; Ribereau-Gayon, 1976); and (-)-hydroxycitrate inhibits the citrate cleavage enzyme (Lowenstein, 1971). As shown in Table 6, all of these compounds are effective inhibitors of fatty acid synthesis by isolated rat hepatocytes. However, in contrast to the inhibition by $Bt_2cAMP$, lactate plus pyruvate fails to reverse the inhibition of fatty acid synthesis imposed by any of these inhibitors (Table 6). Assuming these inhibitors are, indeed, specific for their putative sites of action, they provide a means of dissecting the requirement of lactate plus pyruvate for acetate incorporation into fatty acids. Thus, it should be possible to determine whether this effect is dependent on: (a) pyruvate uptake by the mitochondrion (inhibited by α-cyano-4-hydroxycinnamic acid); (b) transport of the citrate formed *via* intramitochondrial pyruvate metabolism into the cytoplasm (inhibited by 5-tetradecyloxy-2-furoic acid); and (c) cleavage of cytoplasmic citrate to give acetyl-CoA and oxaloacetate (inhibited by (-)-hydroxycitrate). As shown in Table 7, α-cyano-4-hydroxycinnamic acid and 5-tetradecyloxy-2-furoic acid

*Table 7. EFFECT OF CYANOHYDROXYCINNAMIC ACID (CHC), 5-TETRADECYL-OXY-2-FUROIC ACID (TOFA), (-)HYDROXYCITRATE (OHC), AND $Bt_2cAMP$ ON FATTY ACID SYNTHESIS*

| Added substrate | Added inhibitor | Rate of fatty acid synthesis *μmol/min/g* | Relative percent of control rate |
|---|---|---|---|
| $^{14}C$-acetate | None | 0.28 ± 0.04 | 100 |
| | CHC | 0.03 ± 0.01[a] | 11 |
| | TOFA | 0.008 ± 0.002[a] | 3 |
| | OHC | 0.39 ± 0.01[a] | 139 |
| | $Bt_2cAMP$ | 0.038 ± 0.010[a] | 14 |
| | OHC + $Bt_2cAMP$ | 0.11 ± 0.07[a] | 39 |
| | OHC + TOFA | 0.006 ± 0.001[a] | 2 |
| $^{14}C$-acetate + lactate + pyruvate | none | 0.41 ± 0.02 | 100 |
| | CHC | 0.13 ± 0.03[a] | 32 |
| | TOFA | 0.014 ± 0.001[a] | 3 |
| | OHC | 0.65 ± 0.07[a] | 158 |
| | $Bt_2cAMP$ | 0.26 ± 0.02[a] | 63 |
| | OHC + $Bt_2cAMP$ | 0.54 ± 0.03[a] | 132 |

*Rates of fatty acid synthesis were determined between 30 and 60 min of incubation by the addition of $^{14}C$-acetate (10 mM) at 30 min. Initial concentrations were as in Table 6.*

*The results are expressed as mean ± standard error of the mean with three or four rat liver cell preparations in each group. Unpublished results of Harris.*

[a]*$P < 0.05$ by Student's t-test.*

inhibit the synthesis of fatty acids as measured by $^{14}C$-acetate incorporation into fatty acids, regardless of the presence or absence of lactate plus pyruvate. This suggests that the metabolism of pyruvate as well as the generation of cytoplasmic citrate are necessary for $^{14}C$-acetate incorporation. On the other hand, (-)-hydroxycitrate stimulates $^{14}C$-acetate incorporation into fatty acids (Table 7). In addition, lactate plus pyruvate remains effective in reversing $Bt_2cAMP$ inhibition of $^{14}C$-acetate incorporation in the presence of (-)-hydroxycitrate. This suggests that the generation but not the utilization of cytoplasmic citrate is necessary for the effects of lactate plus pyruvate on acetate incorporation. Thus, these results suggest that the maintenance of high levels of cytoplasmic citrate is essential for fatty acid synthesis from cytoplasmic acetyl-CoA. This would be consistent with the well-established allosteric activating effect of citrate upon acetyl-CoA carboxylase. The activating effect of (-)-hydroxycitrate on $^{14}C$-acetate incorporation into fatty acids has been observed by others and explained on the basis of an activation of acetyl-CoA carboxylase by (-)-hydroxycitrate (Barth, Hackenschmidt, Ullmann, and Decker, 1972). However, the $K_a$ for this effect of (-)-hydroxycitrate is much greater than the $K_i$ for inhibition of the citrate cleavage enzyme (Cheema-Dhadli, Halperin, and Leznoff, 1973; Goodridge, 1973b). In addition, inhibition of citrate transport by 5-tetradecyloxy-2-furoic acid completely blocks acetate incorporation in the presence of (-)-hydroxycitrate (Table 7). This argues that citrate transport and the resulting accumulation of cytoplasmic citrate are responsible for (-)-hydroxycitrate stimulation of $^{14}$C-acetate incorporation. The application of the digitonin method (Zuurendonk and Tager, 1974) for measuring cytoplasmic and mitochondrial concentrations of metabolites such as citrate will surely provide more direct evidence for or against this interpretation of the results.

The picture emerging from these studies is that inhibition of lipogenesis by cyclic AMP may be explained, in part, by an inhibition of net glucose utilization, *i.e.*, aerobic glycolysis. This action prevents the accumulation of lactate and pyruvate, causes the inactivation of pyruvate dehydrogenase, and appears from indirect evidence to limit the accumulation of cytoplasmic citrate. The latter may lead to a decrease in the activity of acetyl-CoA carboxylase and an inhibition of fatty acid synthesis. It should be noted, however, that we have not found supplementation with exogenous substrates to completely prevent $Bt_2cAMP$ inhibition of fatty acid synthesis. Thus, additional sites must be involved in the inhibition of fatty acid synthesis caused by cyclic AMP.

## SUMMARY

In 1972 Tepperman and Tepperman reflected that "cyclic AMP appears to bring about nothing less than a deterioration of the total lipogenic climate of the cell." Although some progress has been made, this "not-very-molecular reflection" of the Teppermans remains today the most descriptive statement of the short-term regulatory action of cyclic AMP on hepatic fatty acid synthesis. However, it is important to realize that cyclic AMP is capable of transforming the hepatocyte from a glycolytic and lipogenic mode into a glycogenolytic and gluconeogenic mode. Hence, it is not surprising that many enzymatic steps are involved and that the overall mechanism is complex.

Aerobic glycolysis is clearly important for rapid rates of fatty acid synthesis by isolated hepatocytes. Cyclic AMP, produced either by glucagon activation of adenylate cyclase or from $Bt_2cAMP$, inhibits glycolysis by decreasing the effectiveness of phosphofructokinase and pyruvate kinase. The basis for the decreased effectiveness of phosphofructokinase is not clear but does not appear to involve interconversion to a less active form. Pyruvate kinase is decreased in activity by interconversion to a less active form and by a decrease in the steady state concentration of the positive effector fructose-1,6-diphosphate. The inhibition of glycolysis caused by cyclic AMP decreases the availability of lactate and pyruvate which function not only as substrates but also as part of "pull" mechanism for lipogenesis. Part of the "pull" mechanism involves activation of the pyruvate dehydrogenase complex by inhibition of the kinase which phosphorylates the complex. The "pull" probably also involves feed forward activation of acetyl-CoA carboxylase by citrate. However, the relative importance of citrate in this capacity and the relationship between citrate production and the lactate and pyruvate levels will only be resolved by the determination of its cytoplasmic concentration. The methodology for this is at hand but awaits application to this problem.

Not all of the inhibitory action of cyclic AMP on lipogenesis can be explained by an inhibition of aerobic glycolysis. $Bt_2$cAMP causes significant inhibition of fatty acid synthesis even in the presence of exogenous lactate and pyruvate. This action of cyclic AMP will, perhaps, be explained eventually by an inhibition of citrate transport or the interconversion of acetyl-CoA carboxylase and/or fatty acid synthetase to phosphorylated, inactive forms.

## REFERENCES

Allred, J.B. and Roehrig, K.L. (1973). Inhibition of rat liver acetyl coenzyme A carboxylase by $N^6$, $O^{2'}$-dibutyryl cyclic adenosine 3':5'-monophosphate *in vitro*. *J. Biol. Chem.* 248: 4131-4133.

Allred, J.B. and Roehrig, K.L. (1972). Inhibition of hepatic lipogenesis by cyclic-3', 5'-nucleotide monophosphates. *Biochem. Biophys. Res. Commun.* 46:1135-1139.

Barth, C., Hackenschmidt, J., Ullmann, H., and Decker, K. (1972). Inhibition of cholesterol synthesis by (-)-hydroxy-citrate in perfused rat liver. Evidence for an extramitochondrial mevalonate synthesis from acetyl coenzyme A. *FEBS Lett.* 22:343-346.

Barth, C., Sladek, M., and Decker, K. (1971). The subcellular distribution of short-chain fatty acyl-CoA synthetase activity in rat tissues. *Biochim. Biophys. Acta* 248:24-33.

Batenburg, J.J. and Olson, M.S. (1975). The inactivation of pyruvate dehydrogenase by fatty acid in isolated rat liver mitochondria. *Biochem. Biophys. Res. Commun.* 66:533-540.

Berman, B.G. and Halperin, M.L. (1973). Effect of insulin on pyruvate metabolism in epididymal adipose tissue of the rat: Correlation of intracellular pyruvate contents and pyruvate dehydrogenase activity. *Biochem. J.* 134:885-889.

Blair, J.B., Cimbala, M.A., Foster, J.L., and Morgan, R.A. (1976). Hepatic pyruvate kinase: Regulation by glucagon, cyclic adenosine 3':5'-monophosphate, and insulin in the perfused rat

liver. *J. Biol. Chem.* 251:3756-3762.
Brand, I.A. and Soling, H.D. (1975). Activation and inactivation of rat liver phosphofructokinase by phosphorylation-dephosphorylation. *FEBS Lett.* 57:163-168.
Bremer, J., Bjerve, K.S., Christopherson, B.O., Daae, L.N.W., Solberg, H.E., and Aas, M. (1974). Factors controlling the metabolism of fatty acids in the liver. In: F. Lundquist and N. Tygstrup (Eds.) *Regulation of Hepatic Metabolism*, pp. 159-176. Proceedings of the Alfred Benzon Symposium VI, Munksgaard, Copenhagen.
Bricker, L.A. and Levey, G.S. (1972). Evidence for regulation of cholesterol and fatty acid synthesis in liver by cyclic adenosine 3',5'-monophosphate. *J. Biol. Chem.* 247:4914-4915.
Brunengraber, H., Boutry, M., and Lowenstein, J.M. (1973). Fatty acid and 3 β hydroxysterol synthesis in the perfused rat liver. *J. Biol. Chem.* 248:2656-2669.
Capuzzi, D.M., Rothman, V., and Margolis, S. (1974). The regulation of lipogenesis by cyclic nucleotides in intact hepatocytes prepared by a simplified technique. *J. Biol. Chem.* 249:1286-1294.
Carlson, C.A. and Kim, K.H. (1974). Regulation of hepatic acetyl coenzyme A carboxylase by phosphorylation and dephosphorylation. *Arch. Biochem. Biophys.* 164:478-489.
Cheema-Dhadli, S., Halperin, M.L.,and Leznoff, C.C. (1973). Inhibition of enzymes which interact with citrate by (-) hydroxycitrate and 1,2,3,-tricarboxybenzene. *Eur. J. Biochem.* 38:98-102.
Christiansen, R., Berrebaek, B., and Bremer, J. (1976). The effect of (-) carnitine on the metabolism of palmitate in liver cells isolated from fasted and refed rats. *FEBS Lett.* 62: 313-317.
Clark, D.G., Rognstad, R., and Katz, J. (1974). Lipogenesis in rat hepatocytes. *J. Biol. Chem.* 249:2028-2036.
Claycomb, W.C., Bynagle, J.K., and Kilsheimer, G.S. (1969). Hepatic lipolytic and ketogenic activity in alloxan diabetic rats. *Diabetes* 18(Suppl. 1):350.
Crabb, D.W., Mapes, J.P., Boersma, R.W., and Harris, R.A. (1976). Effect of dichloroacetate on carbohydrate and lipid metabolism of isolated hepatocytes. *Arch. Biochem. Biophys.* 173: 658-665.
Feliu, J.E., Hue, L., and Hers, H-G. (1976). Hormonal control of pyruvate kinase activity and of gluconeogenesis in isolated hepatocytes. *Proc. Nat. Acad. Sci.* USA 73:2762-2766.
Fritz, I.B. (1955). The effects of muscle extracts on the oxidation of palmitic acid by liver slices and homogenates. *Acta Physiol. Scand.* 34:367-385.
Garland, P.B. and Randle, P.J. (1964). Control of pyruvate dehydrogenase in the perfused rat heart by the intracellular concentration of acetyl-coenzyme A. *Biochem. J.* 91:6C-7C.
Geelen, M.J.H. and Gibson, D.M. (1976). In: J.M. Tager, H.D. Soling, and J.R. Williamson (Eds.) *Use of Isolated Liver Cells*

*and Kidney Tubules in Metabolic Studies*, pp. 219-230. North Holland Publishing Co., Amsterdam.

Geelen, M.J.H. and Gibson, D.M. (1975). Lipogenesis in maintenance cultures of rat hepatocytes. *FEBS Lett.* 58:334-339.

Gibson, D.M., Lyons, R.T., Scott, D.F., and Muto, Y. (1972). Synthesis and degradation of the lipogenic enzymes of rat liver. *Adv. Enzym. Reg.* 10:187-204.

Glinsmann, W.H., Hern, E.P., Linarelli, L.G., and Farese, R.U. (1969). Similarities between effects of adenosine 3',5'-monophosphate and guanosine 3',5'-monophosphate on liver and adrenal metabolism. *Endocrinology* 85:711-719.

Goodridge, A.G. (1975). Hormonal regulation of the activity of the fatty acid synthesizing system and of the malic enzyme concentration in liver cells. *Fed. Proc.* 34:117-123.

Goodridge, A.G. (1973a). Regulation of fatty acid synthesis in isolated hepatocytes prepared from the livers of neonatal chicks. *J. Biol. Chem.* 248:1924-1931.

Goodridge, A.G. (1973b). Regulation of fatty acid synthesis in isolated hepatocytes: Evidence for a physiological role for long chain fatty acyl coenzyme A and citrate. *J. Biol. Chem.* 248:4318-4326.

Guynn, R.W., Veloso, D., and Veech, R.L. (1972). The concentration of malonyl-coenzyme A and the control of fatty acid synthesis *in vivo*. *J. Biol. Chem.* 247:7325-7331.

Halestrap, A.P. (1975). The mitochondrial pyruvate carrier: Kinetics and specificity for substrates and inhibitors. *Biochem. J.* 148:85-96.

Halestrap, A.P. and Denton, R.M. (1974). Hormonal regulation of adipose tissue acetyl-coenzyme A carboxylase by changes in the polymeric state of the enzyme: The role of long-chain fatty acyl-coenzyme A thioesters and citrate. *Biochem. J.* 142:365-377.

Halperin, M.L., Cheema-Dhadli, L., Taylor, W.M., and Fritz, I.B. (1974). Role of the citrate transporter in the control of fatty acid synthesis. *Adv. Enzym. Reg.* 13:435-445.

Harris, R.A. (1975). Studies on the inhibition of hepatic lipogenesis by $N^6,O^2$-dibutyryl adenosine 3',5'-monophosphate. *Arch. Biochem. Biophys.* 169:168-180.

Harris, R.A. and Yount, R.A. (1975). Inhibition of hepatic lipogenesis by adenine nucleotides. *Lipids* 10:673-680.

Heimberg, M., Weinstein, I., and Kohout, M. (1969). The effects of glucagon, dibutyryl cyclic adenosine 3',5'-monophosphate, and concentration of free fatty acid on hepatic lipid metabolism. *J. Biol. Chem.* 244:5131-5139.

Kariya, T., Parker, R.A., Grisar, J.M., Martin, J., and Wille, L.J. (1975). Laboratory studies with RMI 14, 514, a new hypolipidemic agent. *Fed. Proc.* 34:789 (Abstract).

Katz, J. and Rognstad, R. (1976). In: B.L. Horecker and E.R. Stadtman (Eds.) *Current Topics in Cellular Regulation*, Vol. 10, pp. 238-290. Academic Press, New York.

Klain, G.J. and Weiser, P.C. (1973). Changes in hepatic fatty acid synthesis following glucagon injections *in vivo*. *Biochem. Biophys. Res. Commun.* 55:76-83.
Lakshmanan, M.R., Nepokroeff, C.M., and Porter, J.W. (1972). Control of the synthesis of fatty-acid synthetase in rat liver by insulin, glucagon, and adenosine 3':5' cyclic monophosphate. *Proc. Nat. Acad. Sci.* USA 69:3516-3519.
Lane, M.D., Moss, J., and Polakis, S.E. (1974). In: B.L. Horecker and E.R. Stadtman (Eds.) *Current Topics in Cellular Regulation*, Vol 8, pp. 139-196.
Lee, K.H., Thrall, T., and Kim, K.H. (1973). Hormonal regulation of acetyl CoA carboxylase. Effect of insulin and epinephrine. *Biochem. Biophys. Res. Commun.* 54:1133-1140.
Linn, T.C., Pettit, F.H., and Reed, L.J. (1969). α-Keto acid dehydrogenase complexes. X Regulation of the activity of the pyruvate dehydrogenase complex from beef kidney mitochondria by phosphorylation and dephosphorylation. *Proc. Nat. Acad. Sci.* USA 62:234-241.
Ljungström, O., Hjelmquist, G., and Engström, L. (1974). Phosphorylation of purified rat liver pyruvate kinase by cyclic 3',5'-AMP-stimulated protein kinase. *Biochim. Biophys. Acta* 358:289-298.
Löffler, G., Bard, S., and Wieland, O.H. (1975). Control of pyruvate dehydrogenase interconversion by palmitoyl-coenzyme A as related to adenine nucleotide translocation in isolated fat cell mitochondria. *FEBS Lett.* 60:269-274.
Lowenstein, J.M. (1971). Effect of (-)-hydroxycitrate on fatty acid synthesis by rat liver *in vivo*. *J. Biol. Chem.* 246:629-632.
Mapes, J.P. and Harris, R.A. (1976). Inhibition of gluconeogenesis and lactate formation from pyruvate by $N^6$, $O^2$-dibutyryl adenosine 3':5'-monophosphate. *J. Biol. Chem.* 251:6189-6196.
Mapes, J.P. and Harris, R.A. (1975). Regulatory function of pyruvate dehydrogenase and the mitochondrion in lipogenesis. *Lipids* 10:757-764.
Mayes, P.A. and Topping, D.L. (1974). Regulation of hepatic lipogenesis by plasma free fatty acids: Simultaneous studies on lipoprotein secretion, cholesterol synthesis, ketogenesis and gluconeogenesis. *Biochem. J.* 140:111-114.
McGarry, J.D. and Foster, D.W. (1974). The metabolism of (-)-octanoylcarnitine in perfused livers from fed and fasted rats. Evidence for a possible regulatory role of carnitine acyltransferase in the control of ketogenesis. *J. Biol. Chem.* 249:7984-7990.
McGarry, J.D., Robles-Valdes, C., and Foster, D.W. (1975a). Role of carnitine in hepatic ketogenesis. *Proc. Nat. Acad. Sci.* USA 72:4385-4388.
McGarry, J.D., Wright, P.H., and Foster, D.W. (1975b). Hormonal control of ketogenesis: Rapid activation of hepatic ketogenic capacity in fed rats by anti-insulin serum and glucagon.

*J. Clin. Invest.* 55:1202-1209.
McGee, R. and Spector, A.A. (1975). Fatty acid biosynthesis in Ehrlich cells: The mechanism of short term control by exogenous fatty acid. *J. Biol. Chem.* 250:5419-5425.
Meikle, A.W., Klain, G.J., and Hannon, J.P. (1973). Inhibition of glucose oxidation and fatty acid synthesis in liver slices from fed, fasted and fasted-refed rats by glucagon, epinephrine and cyclic adenosine-3',5'-monophosphate. *Proc. Soc. Expt. Biol. Med.* 143:379-381.
Metz, R. and Salter, J.M. (1962). Effect of glucagon on liver glycolysis. *Nature* 196:1094-1095.
Nielsen, R.C., Mehlman, M.A., Veech, R.L., and Hawkins, R.A. (1974). Inhibition of fatty acid synthesis by glucagon *in vivo*. *Fed. Proc.* 33:352 (Abstract).
Nilsson, A., Sundler, R., and Akesson, B. (1973). Biosynthesis of fatty acids and cholesterol in isolated rat-liver parenchymal cells: Effect of albumin-bound fatty acids. *Eur. J. Biochem.* 39:613-620.
Nishikori, K., Iritani, N., and Numa, S. (1973). Levels of acetyl coenzyme A carboxylase and its effectors in rat liver after short-term fat-free refeeding. *FEBS Lett.* 32:19-21.
Numa, S. and Yamashita, S. (1974). In: B.L. Horecker and E.R. Stadtman (Eds.) *Current Topics in Cellular Regulation*, Vol. 8, pp. 197-246.
Ochs, R.S. and Harris, R.A. (1977). Differential regulation of pyruvate kinase in rat and chicken liver. *Fed. Proc.* 36:690 (Abstract).
Panek, E., Cook, G.A., and Cornell, N.W. (1976). Personal communication, submitted for publication.
Patzelt, C., Löffler, G., and Wieland, O.H. (1973). Interconversion of pyruvate dehydrogenase in the isolated perfused rat liver. *Eur. J. Biochem.* 33:117-122.
Pettit, F.H., Pelley, J.W., and Reed, L.J. (1975). Regulation of pyruvate dehydrogenase kinase and phosphatase by acetyl-CoA/CoA and NADH/NAD ratios. *Biochem. Biophys. Res. Commun.* 65:575-582.
Qureshi, A.A., Jenik, R.A., Kim, M., Lornitzo, F.A., and Porter, J.W. (1975). Separation of two active forms (holo-*a* and holo-*b*) of pigeon liver fatty acid synthetase and their interconversion by phosphorylation and dephosphorylation. *Biochem. Biophys. Res. Commun.* 66:344-351.
Raskin, P., McGarry, J.D., and Foster, D.W. (1974). Independence of cholesterol and fatty acid biosynthesis from cyclic adenosine monophosphate concentration in the perfused rat liver. *J. Biol. Chem.* 249:6029-6032.
Ribereau-Gayon, G. (1976). Inhibition of mitochondrial tricarboxylate anion translocation and of liver fatty acid synthesis by a new hypolipidemic agent. *FEBS Lett.* 62:309-312.
Salmon, D.M.W., Bowen, N.L., and Hems, D.A. (1974). Synthesis of fatty acids in the perfused mouse liver. *Biochem. J.* 142:

611-618.

Seglen, P.O. (1973). Glycogen synthesis in isolated parenchymal rat liver cells. *FEBS Lett.* 30:25-28.

Severson, D.L., Denton, R.M., Bridges, B.J., and Randle, P.J. (1976). Exchangeable and total calcium pools in mitochondria of rat epididymal fat-pads and isolated fat-cells: Role in the regulation of pyruvate dehydrogenase activity. *Biochem. J.* 154:209-223.

Siess, E.A. and Wieland, O.H. (1976). Phosphorylation state of cytosolic and mitochondrial adenine nucleotides and of pyruvate dehydrogenase in isolated rat liver cells. *Biochem. J.* 156:91-102.

Siess, E.A. and Wieland, O.H. (1975). Regulation of pyruvate dehydrogenase interconversion in isolated hepatocytes by the mitochondrial ATP/ADP ratio. *FEBS Lett.* 52:226-230.

Strandholm, J.J., Cardenas, J.M., and Dyson, R.D. (1975). Pyruvate kinase isozymes in adult and fetal tissues of chicken. *Biochemistry* 14:2242-2252.

Taunton, O.D., Stifel, F.B., Green, H.L., and Herman, R.H. (1974). Rapid reciprocal changes in rat hepatic glycolytic enzyme and fructose diphosphatase activities following insulin and glucagon injection. *J. Biol. Chem.* 249:7228-7239.

Taylor, S.I. and Jungas, R.L. (1974). Regulation of lipogenesis in adipose tissue: The significance of the activation of pyruvate dehydrogenase by insulin. *Arch. Biochem. Biophys.* 164:12-19.

Tepperman, H.M. and Tepperman, J. (1972). Mechanism of inhibition of lipogenesis by 3',5' cyclic AMP. In: I.B. Fritz (Ed.) *Insulin Action.* pp. 543-560. Academic Press, New York.

van Berkel, T.J.C., Kruijt, J.K., Koster, J.F., and Hulsmann, W.C. (1976). Cyclic nucleotide-, pyruvate- and hormone-induced changes in pyruvate kinase activity in isolated rat hepatocytes. *Biochem. Biophys. Res. Commun.* 72:917-925.

Van Tol, A. (1975). Aspects of long-chain acyl-CoA metabolism. *Mol. and Cell. Biochem.* 7:19-31.

Veech, R.L. and Guynn, R.W. (1974). The control of hepatic fatty acid synthesis *in vivo*. In: F. Lundquist and N. Tygstrup (Eds.) *Regulation of Hepatic Metabolism.* pp. 337-357. Proceedings of the Alfred Benzon Symposium VI, Munksgaard, Copenhagen.

Veneziale, C.M. and Swenson, R.P. (1975). Gluconeogenesis in isolated hepatic parenchymal cells. VIII. Effects of monobutyryl cyclic adenosine monophosphate on gluconeogenic intermediates, phosphofructokinase, and fructose diphosphatase. *Mayo Clin. Proc.* 50:271-278.

Watkins, P.A. and Lane, M.D. (1976). Regulation of fatty acid synthesis in chicken liver cells by cAMP mediated control of intracellular citrate levels. *Fed. Proc.* 35:1428 (Abstract).

Walajtys, E.I., Gottesman, D.P., and Williamson, J.R. (1974). Regulation of pyruvate dehydrogenase in rat liver mitochondria by phosphorylation-dephosphorylation. *J. Biol. Chem.* **249**:1857-1865.

Weinstein, I., Klausner, H., and Heimberg, M. (1973). The effect of concentration of glucagon on output of triglyceride, ketone bodies, glucose and urea by the liver. *Biochim. Biophys. Acta* **296**:300-309.

Woods, H.F. and Krebs. H.A. (1971). Lactate production in the perfused rat liver. *Biochem. J.* **125**:129-139.

Zakim, D., Pardini, R.S., Herman, R.H., and Sauberlich, H.E. (1967). Mechanism for the differential effects of high carbohydrate diets on lipogenesis in rat liver. *Biochim. Biophys. Acta* **144**:242-251.

Zuurendonk, P.F. and Tager, J.M. (1974). Rapid separation of particulate components and soluble cytoplasm of isolated rat-liver cells. *Biochim. Biophys. Acta* 333:393-399.

# HORMONAL REGULATION OF LIPOLYSIS: ROLE OF CYCLIC NUCLEOTIDES, ADENOSINE, AND FREE FATTY ACIDS

J.N. Fain and R.E. Shepherd

Section of Physiological Chemistry
Division of Biology and Medicine
Brown University, Box G
Providence, RI 02912

OUTLINE



## Review of the role of cyclic AMP in lipolysis

Twenty years ago, Dole (1956) and Gordon and Cherkes (1956) presented evidence that lipid is mobilized from adipose tissue in the form of free fatty acids noncovalently bound to plasma albumin. The fatty acids are taken up by muscle, liver, and other tissues

---

Our research was supported by NIH grants AM 10149, AM 14868, and NIH postdoctoral fellowship AM 02813 to RES.

where they are either oxidized as a source of energy or reesterified. In addition, free fatty acids (FFA) may also be converted to ketone bodies by liver and, subsequently, oxidized by muscle and the brain as a source of energy.

Free fatty acids are utilized by tissues in proportion to their plasma level. Concentrations of free fatty acids in plasma primarily reflect the net rate of fatty acid release by the fat cells in adipose tissue. This release of fatty acids by fat cells occurs when the rate of fatty acid reesterification is less than the rate of triglyceride hydrolysis. The rate of lipolysis is the major factor regulating the release of FFA since under conditions in which lipolysis is accelerated little α-glycerophosphate is available from glycolysis for fatty acid reesterification. In mammalian fat cells the hydrolysis of the stored triglycerides in adipose tissue is catalyzed by the so-called "hormone-sensitive" lipase which is activated by cyclic AMP. There is another lipase present in adipose tissue which is activated by insulin and referred to as lipoprotein lipase. This lipase is involved in the hydrolysis of plasma lipoproteins and their deposition as triglyceride in adipose tissue. We use the term triglyceride lipase to refer to the enzyme which is activated by lipolytic hormones, inhibited by insulin, and involved in the mobilization of triglyceride stored in adipose tissue.

The activity of triglyceride lipase is directly, or indirectly, enhanced by catecholamines, growth hormone, glucocorticoids, and thyroid hormones. The multiplicity of hormones which regulate lipolysis suggests a prominent role of this process in energy mobilization. Lipolysis can be elevated by either decreasing the level of insulin or increasing the level of a lipolytic hormone. It is still not clear which of the above hormones is responsible for lipid mobilization during fasting, though all are probably involved. The primary hormones believed to regulate lipid mobilization in specific conditions are given below:

| Condition | Hormones |
|---|---|
| Acute mobilization during stress and early onset of exercise | Increased catecholamine release |
| Overnight fast or during prolonged exercise | Elevation in growth hormone and decrease in insulin |
| Diabetic ketoacidosis | Absence of insulin and presence of pituitary hormones |

The lipolytic actions of thyroid hormones have a lag period of four to six hours while those of growth hormone and glucocorticoids have a lag period of approximately one to two hours. In contrast, the lag period for catecholamines is in the order of seconds. Our knowledge of the mechanisms involved in the activa-

*Table 1. AGENTS WHICH INCREASE LIPOLYSIS BY FAT CELLS*

| Agent | Lag period | Is protein synthesis and RNA involved? | What is known about mechanism |
|---|---|---|---|
| Catecholamines Glucagon, ACTH | seconds | No | Activate adenylate cyclase |
| Growth hormone | 1-2 hours | Yes | Increase catecholamine-sensitive adenylate cyclase |
| Glucocorticoids | 1-2 hours | Yes | Does not appear to affect cyclic AMP accumulation |
| Thyroid hormones | 4-6 hours | Not known | Decrease cyclic AMP phosphodiesterase |
| Cholera toxin | 0.5-1 hour | No | Irreversible activation of adenylate cyclase |
| Methylxanthines | seconds | No | Inhibit cyclic AMP phosphodiesterase and adenosine action |
| Adenosine deaminase | seconds | No | Removes adenosine, thus resulting in increased cyclic AMP |

tion of lipolysis by the various hormones is still rather incomplete. Table 1 summarizes the key differences in mechanisms involved for activation of lipolysis by various agents.

Initially our interest in the hormonal regulation of lipolysis was concerned with the role of cyclic AMP in the action of slow-activators of lipolysis such as growth hormone, glucocorticoids, and thyroid hormones. This has turned out to be a difficult problem to address since, even in the case of catecholamines, the involvement of cyclic AMP is not clear. Growth hormone increased glycogen phosphorylase activity and cyclic AMP accumulation by rat fat cells after a lag period of about two hours and these effects were blocked by inhibitors of protein synthesis such as puromycin and cycloheximide (Moskowitz and Fain, 1970). Incubation of fat cells with growth hormone for 3.5 hours prior to the preparation of fat cell ghosts resulted in a 40% increase in the stimulation of ghost adenylate cyclase by norepinephrine (Fain, 1973a), while producing no significant effect on basal, ACTH, or fluoride-stimulated adenylate cyclase. These results suggested that the lipolytic action of growth hormone involved the synthesis of a protein(s) which increased the sensitivity of fat cell adenylate cyclase to activation by catecholamines.

The lipolytic action of glucocorticoids, like that of growth hormone, appears to be permissive since both hormones increase the sensitivity of fat cells to lipolytic agents such as catecholamines. Glucocorticoid addition to fat cells did not affect cyclic AMP accumulation after prolonged incubation either in the absence or presence of growth hormone under conditions in which the latter did increase cyclic AMP (Fain and Loken, 1971; Fain and Saperstein, 1970). The lipolytic action of glucocorticoids was similar to that of growth hormone regarding the nature of the lag period and its sensitivity to inhibitors of RNA and protein synthesis. Lamberts, Timmermans, Kramer-Blankestijn, and Birkenhager (1975) found that cyclic AMP-dependent protein kinase activity increased in homogenates of fat cell suspensions after prior incubation with glucocorticoids. Their data suggest that glucocorticoids might act by potentiating the action of cyclic AMP.

The lipolytic action of catecholamines appears to involve cyclic AMP as based on the following lines of evidence:

1. Catecholamines activate adenylate cyclase in fat cells.
2. Lipolytic agents also activate glycogen phosphorylase in fat cells.
3. Cyclic AMP activates fat cell protein kinase and triglyceride lipase.
4. *Beta* adrenergic antagonists block the elevation of cyclic AMP and of lipolysis due to catecholamines.
5. Adenosine deaminase, cholera toxin, and methylxanthines elevate both lipolysis and cyclic AMP accumulation by fat cells.
6. Adenosine, prostaglandin E, nicotinic acid, and other cyclic carboxylic acids inhibit both cyclic AMP accumulation and lipolysis by fat cells.

The activation of fat cell adenylate cyclase was first reported by Rodbell (1967) using ghosts prepared by lysis of fat cells with hypotonic buffer in the presence of ATP. Birnbaumer and Rodbell (1969) subsequently presented evidence that catecholamines, ACTH, TSH, and glucagon activate the same adenylate cyclase. Fat cell adenylate cyclase preparations currently available are sensitive to agents such as hormones which elevate cyclic AMP in intact cells and long-chain fatty acids which inhibit cyclic AMP accumulation. However, it is not yet possible to obtain effects of prostaglandins, nicotinic acid, lactic acid, or adenosine on adenylate cyclase activity in such broken cell preparations.

Fain, Pointer, and Ward (1972) were able to demonstrate a 22% inhibition of adenylate cyclase activity of rat fat cell ghosts by 73 μM adenosine. In contrast, only 0.2 μM adenosine was required to inhibit cyclic AMP accumulation in intact cells by 50%. Furthermore, as little as 0.05 μM $N^6$(phenylisopropyl) adenosine markedly inhibited cyclic AMP accumulation by intact fat cells but this adenosine analog was completely ineffective as an inhibitor of fat cell ghost adenylate cyclase. The reverse was noted with 2',5'-dideoxyadenosine which was more effective as an inhibitor of adenylate cyclase than of cyclic AMP accumulation by intact cells. These observations suggest that the mechanism by which adenosine inhibits cyclic AMP accumulation in intact cells is quite different from that for inhibition of fat cell ghost adenylate cyclase.

Rodbell (1975) has shown that guanyl-5-yl imidophosphate [GMP-P(NH)P], an analog of GTP resistant to hydrolysis by nucleotide phosphohydrolases, activated fat cell adenylate cyclase at high pH (8.4) and $Mg^{2+}$ (50 *mM*). Rodbell suggested that the regulation of adenylate cyclase is a complex function of the concentrations of both free activating nucleotide, protonated substrate, and the affinities of various transitional states of the enzyme for these ligands.

The presence of a protein kinase in adipose tissue homogenates which is activated by cyclic AMP was demonstrated by Corbin, Reimann, Walsh, and Krebs (1969). Subsequently, it was shown that triglyceride lipase preparations from rat adipose tissue were activated by a cyclic AMP-dependent protein kinase (Khoo, Steinberg, Thompson, and Mayer, 1973). Activation of the lipase was associated with the transfer of label from *gamma* labeled ATP to the enzyme protein. There is no evidence for participation of an intermediate enzyme analogous to muscle phosphorylase b kinase in the activation of triglyceride lipase (Khoo *et al.*, 1973; Khoo, Aguino, and Steinberg, 1974). The best preparations of triglyceride lipase with respect to sensitivity to added cyclic AMP were obtained from chicken adipose tissue by Khoo and Steinberg (1974). The enzyme was activated four- to ten-fold by the addition of muscle protein kinase and cyclic AMP and the effect was reversible upon dialysis.

A major obstacle to assigning cyclic AMP a principal role in mediating catecholamine-involved lipolysis is the lack of correlation between levels of total fat cell cyclic AMP and lipolysis. Fain and Wieser (1975) found that near maximal activation of lipolysis due to catecholamines or theophylline could be observed without detectable rise in cyclic AMP. Fain *et al.* (1972) reported that adenosine could abolish small elevations in cyclic AMP due to either catecholamines or theophylline without affecting lipolysis. Stock and Prilop (1974) reported similar results with prostaglandin $E_1$ and $N^6$(phenylisopropyl) adenosine. Butcher and Baird (1969) found that a rise in cyclic AMP from 0.18 to only 0.40 *nmol/g* resulted in maximal stimulation of lipolysis. Similar results were seen by Fain (1977) in isolated rat fat cells where maximal activation of lipolysis was associated with a rise in cyclic AMP from 0.3 to 0.6 *nmol/g* but cyclic AMP values as high as 6 *nmol/g* were readily obtained. These data support the hypothesis that only small increases in cyclic AMP over basal are necessary for maximal activation of lipolysis. This could occur if there are multiple pools of cyclic AMP. Alternatively an unknown factor could act synergistically with cyclic AMP. In any case, the large increases in cyclic AMP which can be readily observed if fat cells are incubated with lipolytic agents in the presence of methyl xanthines appear to have little physiological significance. It remains to be demonstrated why prostaglandins, adenosine, nicotinic acid, or lactate can abolish the large increases in cyclic AMP due to lipolytic agents without affecting lipolysis (Fain, 1977).

Figure 1 summarizes our current view of the role of cyclic AMP in the hormonal regulation of lipolysis. Catecholamines interact with the *beta*-1 adrenergic receptor and activate adenylate cyclase. Possibly there is involvement of an unknown messenger (X). Adenylate cyclase increases ATP conversion to cyclic AMP which activates protein kinase. The inactive triglyceride lipase is converted to its active form by protein kinase in the presence of ATP, and triglycerides are hydrolyzed to give free fatty acids and glycerol. The free fatty acids act as product inhibitors of the lipase and feedback inhibitors of adenylate cyclase. Adenosine release to the medium is presumably regulated by the level of 5'-AMP. Methylxanthines are shown as inhibitors of adenosine action and of cyclic AMP phosphodiesterase activity. The (-) signs refer to inhibition of adenylate cyclase by adenosine and fatty acids.

## Adenosine and the Regulation of Cyclic AMP Formation in Fat Cells

Adenosine release by rat fat cells. Fain *et al.* (1972) suggested that adenosine might be released to the medium during the incubation of fat cells with lipolytic agents and serve as a feedback regulator of cyclic AMP accumulation. Schwabe, Ebert, and Arbler (1973) shortly thereafter reported that incubated fat cells released an inhibitory substance to the medium which they identified as adenosine. However, they found no increase in adenosine release due to epinephrine. Maximal release of adenosine was seen after 10 *min* incubation and was sustained for the next 20 *min*.

Our attempts at measuring adenosine accumulation in the medium by the specific adenosine kinase assay of Namm and Leader (1974)

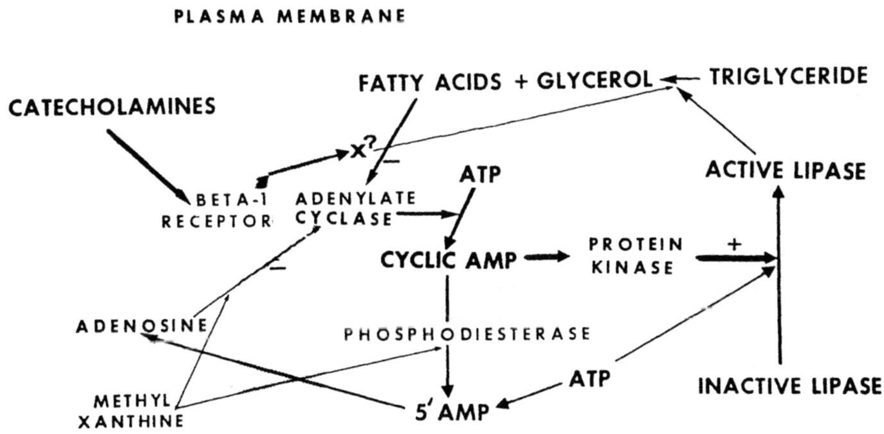


*Figure 1. 1976 model for activation of lipolysis by catecholamines and its modulation by methyl xanthines, fatty acids, and adenosine.*

were unsuccessful. In preliminary studies done in collaboration with Dr. D. Namm of Burroughs Wellcome, Raleigh, NC, we found that recovery from the medium of exogenously added adenosine was poor. These findings suggested that the albumin in the medium contained adenosine deaminase activity. We confirmed this by demonstrating that the Armour bovine fraction V albumin preparations used for our studies deaminated 36% of the added adenosine (0.1 μM) to inosine over 10 *min* at an albumin concentration of 3%.

It was possible to see a transient accumulation of adenosine in the 3% albumin medium if a large number of fat cells were incubated in a small volume of the medium (Figure 2). Maximal accumulation of adenosine occurred at 10 *min*; by 30 *min* all the adenosine had disappeared. Surprisingly, this accumulation of adenosine in the medium was not altered by lipolytic agents such as glucagon or theophylline (Figure 2).

The measurements of adenosine release shown in Figure 2 were based on the accumulation of tritiated adenosine derived from labeled ATP in fat cells. The ATP was labeled by incubating adipose tissue with adenine (2-$^3$H) during the collagenase digestion of adipose tissue. This procedure was originally used to measure labeled cyclic AMP formation from ATP (Fain, 1973a) and is based on the assumption that the entire pool of ATP is labeled. Furthermore, the assumption must be made that the specific activity of the precursors of adenosine in the medium are not altered by hormones or other agents. Measurements of total cyclic AMP compared to radioactive cyclic AMP derived from labeled ATP by prior incubation of fat cells with adenine (2-$^3$H) give similar results

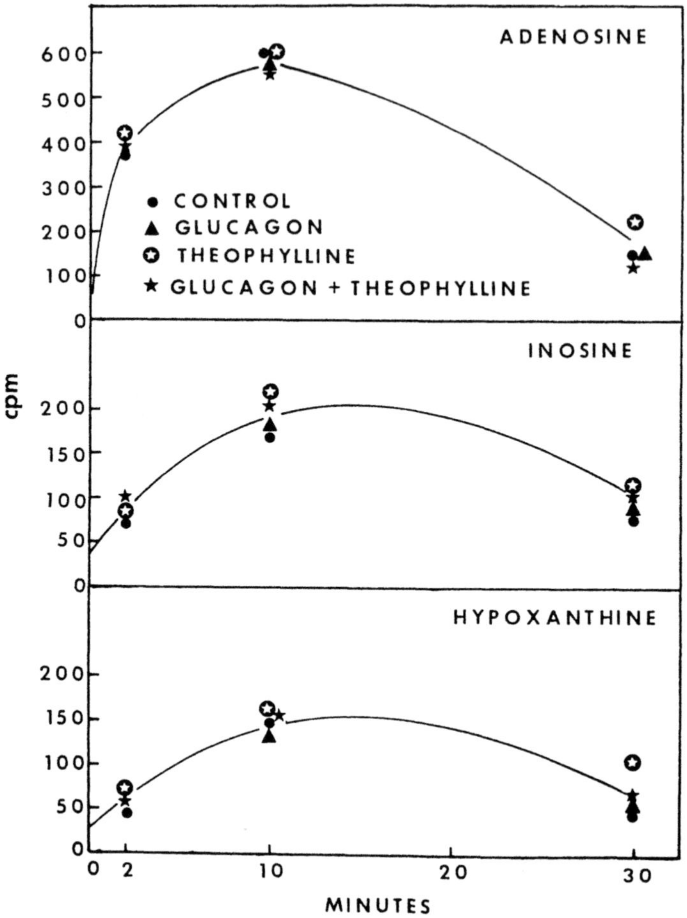


*Figure 2. Influence of glucagon and theophylline on the appearance of labeled inosine, adenosine, and hypoxanthine in the medium after incubation of fat cells with tritiated adenine. Adipose tissue was incubated with 100 μc of adenine (2-$^3$H) and collagenase for 45 min. The fat cells were isolated and then 2.5 ml of packed fat cells were incubated for 20 min with 10 ml of medium without any label. The cells were then washed twice and 0.6 ml of packed cells (450 mg) were added to 2 ml of medium in the absence or presence of 1 μg/ml of glucagon, 200 μM theophylline, or both. Fifty μl aliquots of the medium were removed at the times indicated and added to tubes containing 5 μl of 4N $HClO_4$. Aliquots (5 μl) of the perchloric acid extracts obtained after centrifugation were spotted on PEI cellulose chromatography plates and ascending chromatography using glacial acetic acid, butanol, and water (25:50:25) was carried out. The amount of radioactivity present in fat cell nucleosides and nucleotides was also determined at the end of the 30 min incubation. Most of the label was present in ATP and the percentage of fat cell radioactivity present in ATP and the percentage of fat cell radioactivity present as adenosine, inosine and hypoxanthine was 2.6%, 1.5%, and 0.9%. The total amount of label present in the medium at 30 min was less than 3% of that present in the cells and adenosine, inosine, and hypoxanthine accounted for 35%, 20%, and 14% of the medium label. There was virtually no labeled adenine present in either the cells or the medium (cpm present in either was less than 10 at all times).*

(Fain, 1971; Fain and Loken, 1971; Fain *et al.*, 1972). We used this methodology to measure labeled adenosine release since other procedures were far too insensitive.

The disappearance of adenosine from the medium between 10 and 30 *min* of incubation could not be accounted for by a net increase in inosine or hypoxanthine (Figure 2). We added adenosine-8-$^{14}$C at a concentration of 0.6 µM to fat cells incubated in 3% albumin buffer and found that some was incorporated into fat cell nucleotides but most were oxidized to uric acid (Table 2). In the presence of erythro-9-(2-hydroxy-3-nonyl) adenine (EHNA) which is an inhibitor of adenosine deaminase (Schaeffer and Schwender, 1974) the disappearance of adenosine was markedly reduced, inosine formation was virtually abolished and uric acid formation was reduced by 66% (Table 2). This suggested that the rate-limiting step in uric acid formation is the deamination of adenosine to inosine. In

*Table 2. COMPARISON OF ADENOSINE METABOLISM BY RAT AND CHICKEN FAT CELLS*

| Additions | Time in min | Nucleo-tides | Adeno-sine | Inosine | Hypo-xanthine | Uric acid |
|---|---|---|---|---|---|---|
| | | | Rat Fat Cells | | | |
| None | 0 | <1 | 404 | 80 | 7 | 9 |
| None | 30 | 91 | 41 | 61 | 37 | 227 |
| NBTG 10 µM | 30 | 74 | 60 | 76 | 31 | 214 |
| EHNA, 2 µM | 30 | 122 | 266 | 7 | 7 | 86 |
| EHNA + NBTG, 10 µM | 30 | 82 | 268 | 9 | 10 | 87 |
| | | | Chicken Fat Cells | | | |
| None | 0 | <1 | 232 | 226 | 7 | 4 |
| None | 30 | 106 | 10 | 280 | 75 | 5 |
| NBTG, 10 µM | 30 | 7 | 40 | 405 | 36 | 3 |
| EHNA, 2 µM | 30 | 205 | 74 | 156 | 48 | 2 |
| EHNA + NBTG, 10 µM | 30 | 9 | 243 | 205 | 32 | 10 |

*Rat (63 mg/mℓ) or chicken (67 mg/mℓ) fat cells were incubated for 30 min in the presence of 0.6 µM adenosine-8-$^{14}$C. The values represent the cpm present in 10 µℓ of a perchloric acid extract of cells plus medium. EHNA is the abbreviation for the adenosine deaminase inhibitor erythro-9-(2-hydroxy-3-nonyl) adenine and NBTG for p-nitrobenzylthioguanosine.*

*Table 3. DISTRIBUTION OF LABEL FROM ATP IN CELLS AND MEDIUM IN RAT FAT CELLS AFTER PRIOR INCUBATION WITH ADENINE-8-$^{14}$C*

| Area on thin-layer chromatograph | 10 minutes cells | 10 minutes medium | 60 minutes cells | 60 minutes medium | 60 minutes + valinomycin 1 μg/mℓ cells | 60 minutes + valinomycin 1 μg/mℓ medium |
|---|---|---|---|---|---|---|
| | % distribution of label | | | | | |
| ATP, ADP | 93 | 26 | 92 | 19 | 35 | 4 |
| AMP | 2.5 | 24 | 6 | 4 | 63 | 5 |
| Inosine, xanthine hypoxanthine | 2.0 | 10 | <1 | 2 | <1 | 7 |
| Adenosine | <1 | 31 | <1 | 3 | <1 | 4 |
| Uric acid | 2.5 | 10 | 1 | 75 | 1 | 80 |
| Total label present in cpm | 1260 | 147 | 1031 | 198 | 886 | 404 |

*Rat fat cells (75 mg/1.2 mℓ) were incubated for 10 or 60 min in medium containing 3% Armour albumin. The cells had been labeled with adenine-8-$^{14}$C during digestion of 4 g of adipose tissue in 8 mℓ of 3% albumin buffer containing 4 mg of collagenase and 25 μc of adenine-8-$^{14}$C (0.50 μmoℓ).*

of adenine-2-$^{3}$H labeled ATP to water was enhanced by valinomycin. In the oxidation of hypoxanthine to uric acid the tritium on the 2-position of the purine moiety is removed and appears as water. This observation prompted us to examine the possibility that valinomycin increased the conversion of ATP to uric acid. Valinomycin the presence of allopurinol, which is a potent inhibitor of xanthine dehydrogenase, uric acid formation was abolished and hypoxanthine formation increased (unpublished studies).

Further studies on adenosine release were done using cells in which the ATP had been labeled with adenine-8-$^{14}$C so that the accumulation of uric acid could be measured. The data in Table 3 indicate that over 90% of the label present in fat cells, even after 60 *min* incubation, was present in ATP and ADP. At 10 *min* there was little ATP or ADP present in the medium but there were substantial amounts of AMP and adenosine. Most of the label present as AMP and adenosine at 10 *min* was converted to uric acid by the end of the 60 *min* incubation (Table 3).

Fain and Loken (1971) noted that valinomycin, an antibiotic which increases the permeability of membranes to potassium, inhibited fat cell lipolysis. They further noted that the conversion

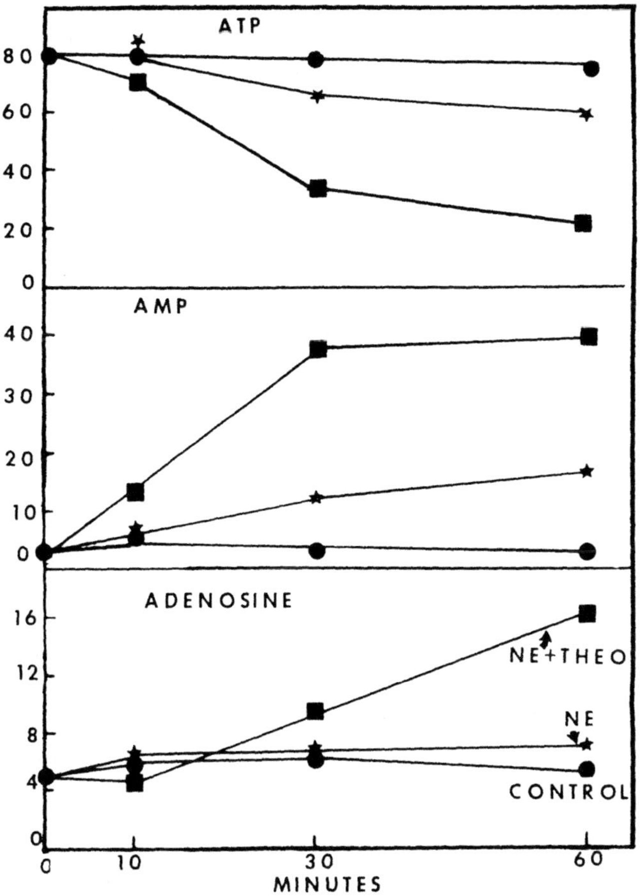


*Figure 3. Stimulation by lipolytic agents of adenosine accumulation in the presence of an inhibitor of adenosine deaminase. Rat fat cells (50/mg/mℓ) were incubated for the times indicated after the nucleotide pool had been labeled by incubation with adenine-8-$^{14}C$. The filled circles represent control values, the stars those for cells incubated in the presence of 1.5 μM norepinephrine and the filled squares those for norepinephrine in the presence of 200 μM theophylline. The cells were incubated in the presence of 2 μM erythro-9-(2-hydroxy-3-nonyl) adenine, an inhibitor of adenosine deaminase. The values represent the sum of metabolites present in cells plus medium and are shown as percentage distribution of total label in ATP, AMP, and adenosine. There was less than 1% accumulation of label in inosine, xanthine, or hypoxanthine. About 11% of the label was initially present in uric acid and this increased to 21% after 60 min incubation. Glycerol release was 7 and 11.0 μmoℓ/g after 30 and 60 min of incubation in the presence of norepinephrine alone and 10.0 and 13.2 μmoℓ/g at 30 and 60 min in the presence of norepinephrine plus theophylline. Glycerol release in the absence of lipolytic agents was less than 0.1 μmoℓ/g.*

caused a marked increase in uric acid formation from ATP that was accompanied by a shift in fat cell label from ATP to AMP (Table 3).

We next examined the effects of agents on adenosine release in the presence of an inhibitor of adenosine deaminase. The data in Figure 3 indicate that even in the presence of EHNA there was no net accumulation of adenosine over an hour incubation except in the presence of norepinephrine plus theophylline. Though norepinephrine alone maximally stimulated lipolysis and produced a small rise in labeled AMP at 30 and 60 *min*, no increase in adenosine release was noted. Increases in adenosine release due to norepinephrine plus theophylline were probably secondary to the large shift in label from ATP to AMP thereby increasing the availability of substrate for the 5-nucleotidase.

Similar results were noted in the studies shown in Figure 4 where an inhibitor of xanthine dehydrogenase was also present. Under these conditions adenosine and hypoxanthine accumulate in the medium. We have been unable to see any increase in adenosine

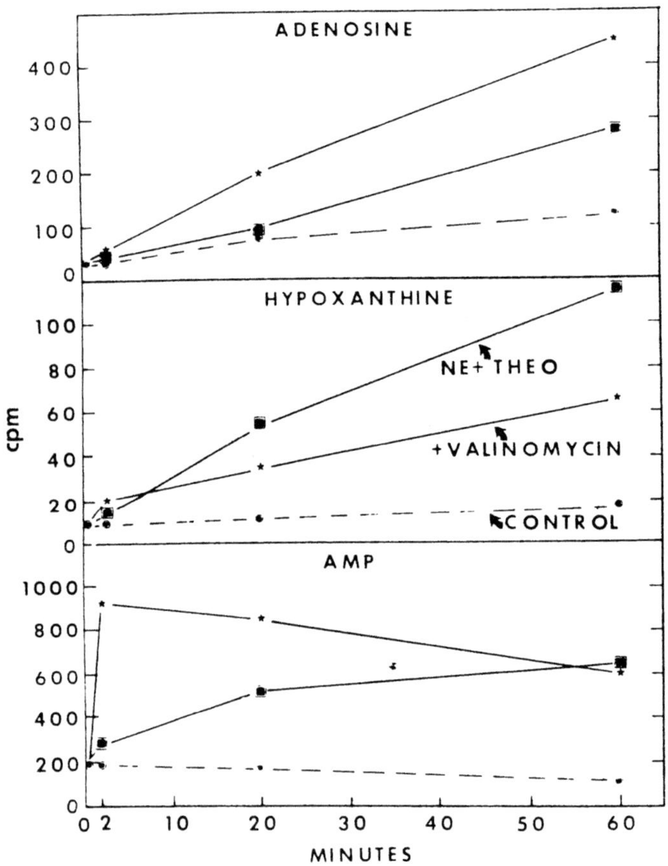


*Figure 4. Appearance of label derived from ATP in AMP, adenosine, and hypoxanthine during incubation of rat fat cells. Rat fat cells (55 mg/mℓ) were incubated for the times indicated after the nucleotide pool had been labeled by incubation with adenine-2-$^3H$. The fat cells were incubated and isolated in medium containing 10 μM allopurinol and 2 mM erythro-9(2-hydroxy-3-nonyl) adenine. The pooled fat cells were incubated in buffer without labeled adenine for 10 min, then rinsed with fresh buffer and incubated for the times shown in the absence or presence of 1 μg/mℓ of valinomycin or 1.5 μM norepinephrine plus 200 μM theophylline. The values are for cells plus medium.*

release due to hormones without a concomitant increase in labeled AMP (Figures 3 and 4). The addition of norepinephrine in the presence of theophylline increased labeled AMP at 20 *min* and adenosine release at 60 *min*. Addition of valinomycin caused a near quantitative conversion of labeled ATP to AMP within two *min* and a marked increase in adenosine release at 20 *min* (Figure 4).

Our conclusion is that the release of adenosine is enhanced only under conditions in which fat cell AMP is elevated. Angel, Desai, and Halperin (1971) have suggested that the drop in ATP and rise in AMP due to lipolytic agents is secondary to an intracellular accumulation of free fatty acids. They suggested that this occurs only in the presence of high concentrations of lipolytic agents if the primary binding sites on the medium albumin are saturated.

The pathways for adenosine metabolism are summarized in Figure 5. The major reactions for formation and metabolism of adenosine are shown by the heavy lines. The pathway for adenylate metabolism via adenylate deaminase is a possible alternate pathway but no evidence is available which suggests that it is of any significance in rat fat cells.

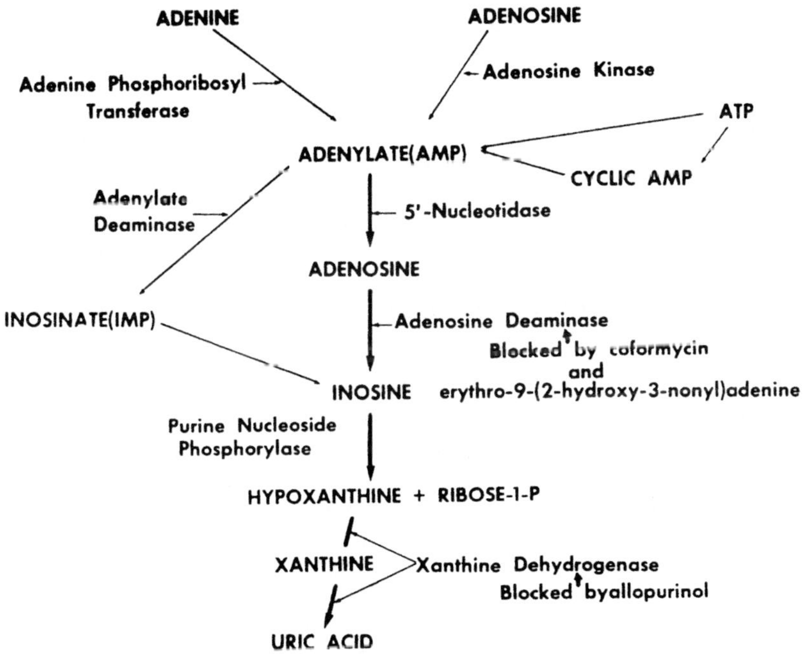


*Figure 5. Pathways for adenosine metabolism in rat fat cells.*

Possible mechanisms for adenosine inhibition of cyclic AMP accumulation. One hypothesis is that adenosine is an inhibitor of adenylate cyclase and binds to a regulatory site on the plasma membrane. We have evidence from a variety of experiments suggesting that neither metabolism nor uptake of adenosine into fat cells is required for inhibition of cyclic AMP accumulation.

Cyclic AMP accumulation in fat cells is inhibited by 2'5'-dideoxyadenosine, which cannot be phosphorylated to nucleotides, and by $N^6$-(phenylsopropyl) adenosine, which cannot be deaminated to inosine (Fain, 1973b). Neither papaverine nor dipyridamole, inhibitors of adenosine uptake in many cells, were able to block the decrease in cyclic AMP accumulation due to adenosine (Fain, 1973b).

Paterson and Oliver (1971) reported that p-nitrobenzylthioguanosine (NBTG) was a potent inhibitor of adenosine uptake and incorporation into nucleotides by many cells. However, the data shown in Table 2 indicate that this nucleoside markedly inhibits adenosine incorporation into nucleotides by chicken but not rat fat cells. We find no inhibitory effect of 10 μM NBTG on the log-dose response curve for inhibition by adenosine of cyclic AMP accumulation in chicken fat cells (unpublished experiments). Our results are similar to those of Haslam and Rosson (1975) who found that NBTG inhibited adenosine incorporation into nucleotides in platelets without affecting the stimulation of cyclic AMP accumulation by adenosine.

Further support for the hypothesis that adenosine may inhibit cyclic AMP accumulation by binding to a site which modifies cyclic AMP metabolism comes from studies with adenosine covalently attached to stachyose. Olsson, Davis, Khouri, and Patterson (1976) found that approximately 2.95 *moles* of adenosine can be coupled per *mole* of stachyose resulting in a water soluble derivative (STADO). They observed that the adenosine linked to stachyose (STADO) is able to produce a dose-dependent coronary vasodilation after intracoronary infusion into dogs.

In chicken fat cells STADO is more effective than adenosine in inhibiting cyclic AMP accumulation due to glucagon during a five *min* incubation (Figure 6). Since STADO cannot be inactivated by deamination, its activity was compared to adenosine in the presence of EHNA, an inhibitor of adenosine deaminase. Under these conditions adenosine and STADO were equipotent. In rat fat cells STADO was equipotent to adenosine in the absence of EHNA but less potent than adenosine in the presence of EHNA (Figure 6). STADO also inhibited the rise in cyclic AMP due to norepinephrine + theophylline equally well during a one-*min* incubation (unpublished results).

We suggest that adenosine inhibits cyclic AMP accumulation by

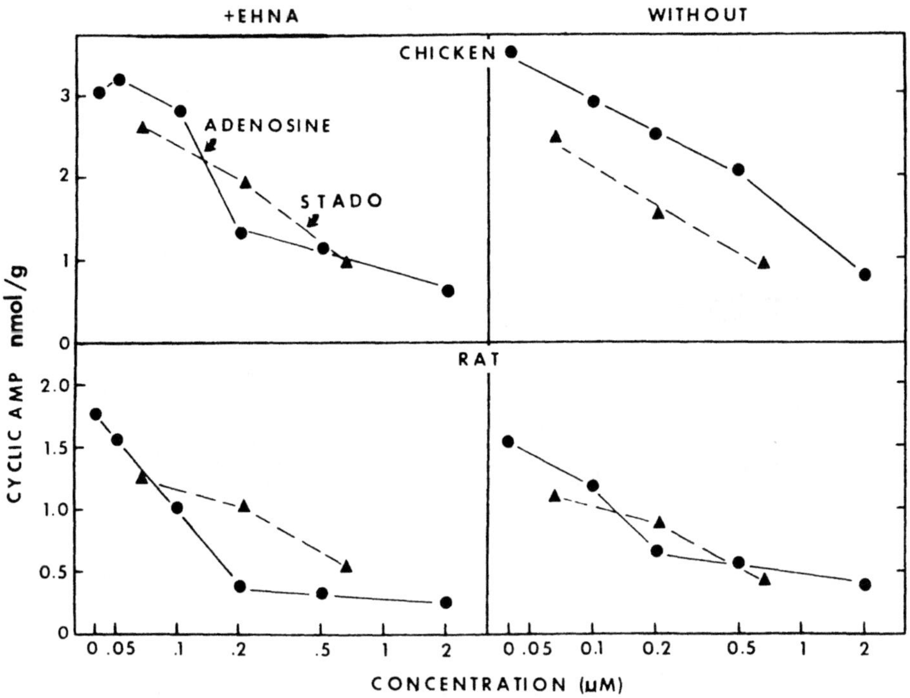


*Figure 6.* *Comparison of inhibition of cyclic AMP accumulation by adenosine and STADO in rat as compared to chicken fat cells. Rat (46 mg/mℓ) or chicken (32 mg/mℓ) fat cells were incubated for 5 min in the presence of 1 μg/mℓ of glucagon. Theophylline (100 μM) was also present during the incubation of rat fat cells. Cells were incubated either with (left side of figure) or without (right side of figure) 2 μM EHNA (erythro-9-(2-hydroxy-3-nonyl) adenine. The circles represent cells incubated without or with varying concentrations of adenosine and the triangles represent incubated cells with varying amounts of STADO.*

*The stachyose-iminobis (propylamino) purine riboside (STADO) was prepared by reacting 3-(9-β-D-ribofuranosyl-purin-6-yl)-3'-iminobis (propylamine) with the tetra-aldehyde obtained by periodate oxidation of digalactosyl sucrose (stachyose) as described by Olsson et al. (1976). The schiff base which resulted was reduced with sodium borohydride to give STADO. Approximately 3 moles of the adenosine analog generally reacted with each mole of oxidized stachyose. The concentration of the STADO is expressed in this paper in molar equivalents of stachyose and assumes that only one of the three adenosine analogs linked to each mole of stachyose is active.*

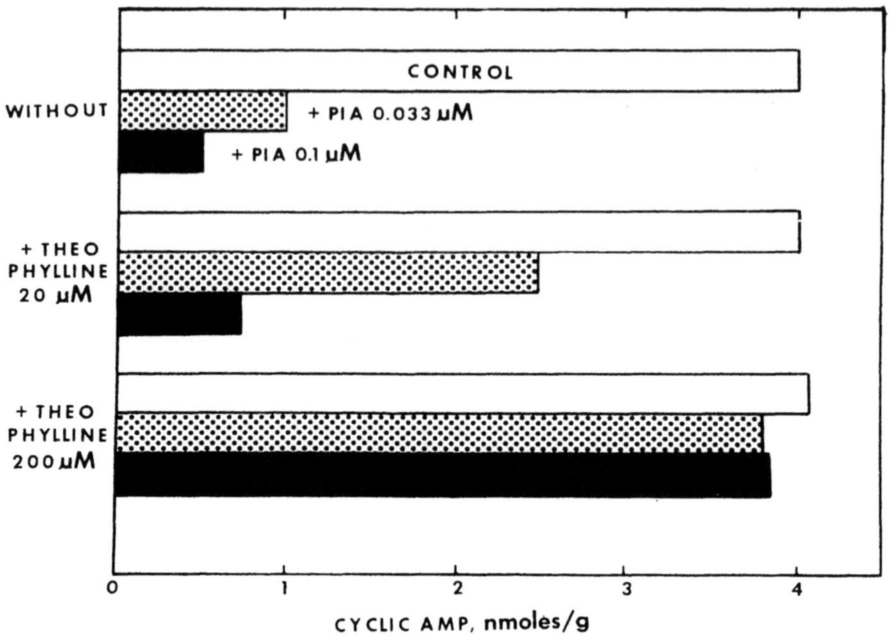


*Figure 7.* *Reversal of* $N^6$*(phenylisopropyl) adenosine action by theophylline. Chicken fat cells (20 mg/tube) were incubated for 30 min in the presence of 50 ng/mℓ of glucagon. PIA represent* $N^6$*(phenylisopropyl) adenosine which was added at a concentration of 0.33* μM *(stippled bars) or 0.1* μM *(solid bars) in the absence or presence of theophylline. The values are the means of two paired experiments.*

interacting directly with a site on the plasma membrane of fat cells that regulates adenylate cyclase since STADO was virtually as potent as adenosine was in inhibiting cyclic AMP accumulation by fat cells (Figure 3). The possibility that adenosine was split from the stachyose during incubation with fat cells seems unlikely. The adenosine derivative was virtually as active as adenosine and the effects were just as pronounced after a 60-second incubation with fat cells as after longer periods.

Methylxanthines and adenosine interactions. Kitabgi (1976) and Malgieri, Shepherd, and Fain (1975) demonstrated that theophylline did not potentiate the rise in cyclic AMP accumulation by chicken fat cells incubated with glucagon. In contrast, theophylline markedly increased the rise in cyclic AMP due to glucagon addition to rat fat cells. The data in Figure 7 indicate that chicken fat cells were responsive to 0.033 or 0.1 μM $N^5$ (phenylisopropyl) adenosine (PIA), an adenosine analog which cannot be deaminated. Theophylline reversed the inhibition of cyclic AMP accumulation due to the adenosine analog (Figure 7). In the absence of theophylline, 3.3 μM adenosine inhibited the glucagon stimulated rise in cyclic AMP accumulation by chicken fat cells by 75%. This concentration of adenosine inhibited cyclic AMP by only 10% in the presence of 100 μM theophylline (unpublished studies). These data suggest that theophylline acts by antagonizing the inhibition of fat cell adenylate cyclase by endogenous adenosine. Since chicken fat cells have little, if any, adenosine present, glucagon alone is able to maximally activate cyclic AMP accumulation and theophylline has no effect unless exogenous adenosine is added.

Previous studies with fat cells isolated from newly hatched chickens indicated that there was no effect on either lipolysis or cyclic AMP accumulation when adenosine deaminase was added to the medium (Boyd, Weiser, and Fain, 1975). These data suggest that chicken fat cells can respond to adenosine but ordinarily there is very little adenosine present. Additionally, chicken fat cells deaminate much more adenosine to inosine than is the case for rat fat cells (Table 2) and accumulate less adenosine in the medium in response to lipolytic agents than do rat fat cells (Figure 8). Even in the presence of an inhibitor of adenosine deaminase, adenosine is not released by chicken fat cells while there is appreciable release by rat fat cells. There was some release of labeled adenosine if 200 μM adenosine was added to the medium. This release may represent either inhibition of reutilization of labeled adenosine or increased turnover to ATP in the presence of large amounts of adenosine. There was an increase in uric acid formation in rat fat cells in the presence of added adenosine which suggests an increased turnover of ATP. The failure of exogenous adenosine to decrease the appearance of label in inosine or uric acid indicates that their formation from labeled nucleotides does not involve a pool of adenosine that equilibrates with adenosine in the medium.

The data in Figure 8 and Table 2 indicate that chicken fat cells do not convert labeled ATP or adenosine to uric acid. The reason for this difference between chicken and rat fat cells is not clear. We have been unable to find any effect of allopurinol on rat fat cell lipolysis (unpublished experiments) under conditions where xanthine dehydrogenase activity is blocked.

## Free Fatty Acids as Feedback Regulator of Adenylate Cyclase

Apparently the major feedback regulator of cyclic AMP accumulation released during incubation of rat fat cells with lipolytic agents is not adenosine. Schwabe and Ebert (1974) found that even in the presence of adenosine deaminase the rise in cyclic AMP due to norepinephrine plus theophylline returned to basal values by 60 minutes. Fain and Shepherd (1975) found a potent inhibitor of cyclic AMP accumulation whose release to the medium was seen only in

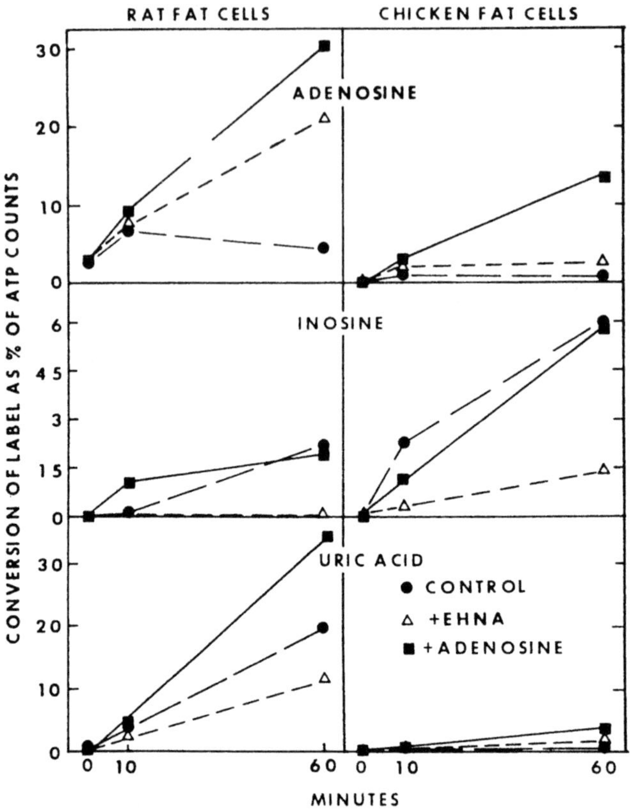


*Figure 8. Accumulation of uric acid, inosine, and adenosine by rat and chicken fat cells. Rat (70 mg/mℓ) or chicken (50 mg/mℓ) fat cells were incubated for the times indicated. The nucleotide pools of the fat cells had been labeled by incubation with adenine-8-$^{14}C$ during digestion of the tissue with collagenase. The data are shown as percent conversion of label initially present in ATP or adenosine, inosine, or hypoxanthine in cells plus medium in the absence (open triangles) or presence of 200 µM adenosine (filled squares) or 2 µM EHNA (erythro-9-(2-hydroxy-3-nonyl) adenine (filled circles) an inhibitor of adenosine deaminase.*

the presence of lipolytic agents and could not be reversed by adenosine deaminase. This inhibitor was also differentiated from adenosine in that it was nondialyzable.

Ho and Sutherland (1971) and Manganiello, Murad, and Vaughan (1971) reported that incubation of rat fat cells with lipolytic agents resulted in the formation of antagonists that blocked the ability of cells to elevate cyclic AMP in response to a second addition of hormone. The formation of the antagonists was enhanced by incubation with lipolytic agents such as epinephrine, ACTH, glucagon, or dibutyryl 3'5' cyclic AMP. The antagonists which accumulated in the presence of epinephrine also blocked the subsequent response to ACTH and glucagon.

The response of cells which had been incubated with lipolytic agents was restored by washing the cells and incubating them in fresh buffer (Ho and Sutherland, 1971; Manganiello *et al.*, 1971). This suggested that the antagonists were released to the medium. Subsequent work has confirmed that adenylate cyclase inhibitors are released to the medium by rat fat cells incubated with lipolytic agents (Fain and Shepherd, 1975). The antagonists do not act by increasing cyclic AMP phosphodiesterase activity (Pawlson, Lovell-Smith, Manganiello, and Vaughan, 1974) in agreement with the initial report of Manganiello *et al.* (1971).

Illiano and Cuatrecasas (1971) postulated that prostaglandins were physiologically important feedback regulators of lipolysis based on studies with indomethacin. This drug has been shown to effectively block prostaglandin formation in rat fat cells by Dalton and Hope (1974). Dalton and Hope (1973); Fain, Psychoyos, Czernik, Frost, and Cash (1973); and Fredholm and Hedqvist (1975) reported that they were unable to see any potentiation of lipolysis or cyclic AMP accumulation by indomethacin in studies with rat fat cells. Their data do not substantiate the hypothesis that prostaglandins are important feedback regulators of lipolysis and cyclic AMP accumulation by rat fat cells.

Gorman (1975) suggested that prostaglandin endoperoxide intermediates, rather than prostaglandins, act as feedback regulators of cyclic AMP accumulation. Gorman, Hamberg, and Samuelson (1975) reported that $PGH_2$ (15-hydroxy-9-peroxidoprosta-5,13-dienoic acid) at 0.25 to 0.28 μM inhibited adenylate cyclase activity of fat cell ghosts. The real question is whether indomethacin also inhibits the formation of prostaglandin endoperoxides. In tissues that have been tested, indomethacin blocked endoperoxide formation as well as that of prostaglandins (Hamberg and Samuelson, 1974). Furthermore, endoperoxides in platelets have functions just the opposite of those for E series prostaglandins.

During the activation of lipolysis by hormones such as catecholamines and ACTH, the fatty acids released to the medium appear to be the major feedback regulators of adenylate cyclase. The important factor is the ratio of free fatty acid to albumin in the medium. Rodbell (1965) first demonstrated that lipolysis by isolated fat cells in the presence of hormones virtually ceased when the ratio of free fatty acids to albumin exceeds 3. Only recently have we realized that if the primary binding sites on medium albumin are saturated with free fatty acids there is an inhibition of both adenylate cyclase and triglyceride lipase.

The first report to show inhibition of cyclic AMP accumulation by free fatty acids was that of Burns, Langley, and Robison (1975). They incubated human fat cells with medium containing enough added sodium oleate to give a free fatty acid to albumin ratio of 2.5 to 3 and found marked inhibition of cyclic AMP accumulation. There was little stimulation of cyclic AMP accumulation in human fat cells by lipolytic agents unless albumin was present in the medium. Unlike rat fat cells (Fain and Shepherd, 1975) there was no direct inhibition of human fat cell adenylate cyclase by free fatty acids (Burns *et al.*, 1975). However, in ghosts prepared from human fat cells previously incubated with high fatty acid to albumin ratios, there was a reduced sensitivity to epinephrine but not to fluoride (Burns *et al.*, 1975).

Medium in which rat fat cells had previously been incubated with lipolytic agents, or with added free fatty acids, markedly inhibited cyclic AMP accumulation by a fresh batch of fat cells (Fain and Shepherd, 1975). The addition of sodium oleate to the medium in amounts sufficient to raise the free fatty acid to albumin ratio above 2 also inhibited cyclic AMP accumulation with the maximal inhibition being seen at a ratio of 5.

The inhibition of cyclic AMP accumulation by fatty acid/albumin ratios above 2 was rapid in onset with near maximal effects being seen after only 30 *sec* (Fain and Shepherd, 1975). This suggested that the fatty acids directly inhibited adenylate cyclase. Fain and Shepherd (1975) reported a direct inhibition of hormone-sensitive adenylate cyclase activity of rat fat cell ghosts by dialyzed medium which had previously been incubated with fat cells in the presence of lipolytic agents. The effects of the medium could be mimicked by the addition of oleate to fat cell ghosts in amounts sufficient to raise the fatty acid/albumin ratio above 2 and maximal inhibition was seen at a ratio of 6 (Fain and Shepherd, 1975).

Inhibition of adenylate cyclase by free fatty acid/albumin ratios of 2.5 or 4.5 was seen at all concentrations of norepinephrine, glucagon, or fluoride (Figure 9). These data indicate that the inhibitory effect of oleate cannot be reversed by increasing the concentration of hormone or fluoride. The effect of a free fatty acid/albumin ratio of 2.5 was slight but that of a ratio of 4.5 was quite marked and also seen with respect to basal cyclase activity. The inhibitory effect of oleate on norepinephrine activation of cyclase was partially reversed by increasing the $Mg^{2+}$ concentration in the assay from 5 to 20 *mM* (unpublished studies). The stimulation due to norepinephrine alone was unaffected by increasing $Mg^{2+}$.

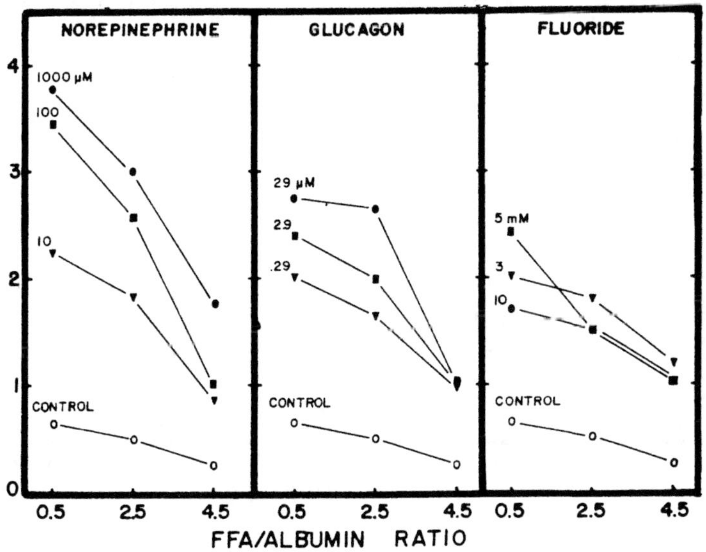


*Figure 9. Inhibition of adenylate cyclase by oleate. Rat fat cell ghosts (38 µg protein per tube) were incubated without or with norepinephrine, glucagon, or sodium fluoride in the absence or presence of oleate. Oleate was incubated with albumin to give final oleate/albumin ratios of 0.5, 2.5, and 4.5. The concentration of albumin was 450 µM. Incubations were done at 37°C for 10 min. Ghosts were prepared by hypotonic lysis of isolated fat cells at room temperature and were used immediately following preparation. The lysing medium contained 5 mM $MgCl_2$, 2.5 mM ATP, 1 mM $KHCO_3$, 2 mM Tris (pH 8.0), 30 mM KCl, 5 mM $MgCl_2$, 1 mM ATP, 8 mM creatine phosphate, and 0.5 U/mℓ creatine phosphokinase. Concentrations of norepinephrine (µM), glucagon (µM), and sodium fluoride (mM) are given in the figure above the corresponding data points for values at an FFA/albumin ratio of 0.5. Adenylate cyclase activity is given as nanomoles of cyclic AMP formed per mg protein. The values are the means of three paired experiments.*

Rodbell (1975) demonstrated that 5'-guanylylimido-diphosphate (GMP-P(NH)P) is a potent activator of adenylate cyclase in fat cells. The increase in cyclase activity due to GMP-P(NH)P is slow in onset and irreversible after prolonged incubation. We found that oleate enhanced the activation of adenylate cyclase by GMP-P(NH)P under conditions in which it inhibited that due to norepinephrine and had little effect on that due to fluoride (Figure 10). The most striking differences were seen during the first 5 *min* of incubation and oleate reduced the lag period required before cyclase was activated by the guanine nucleotide. Possibly oleate interferes

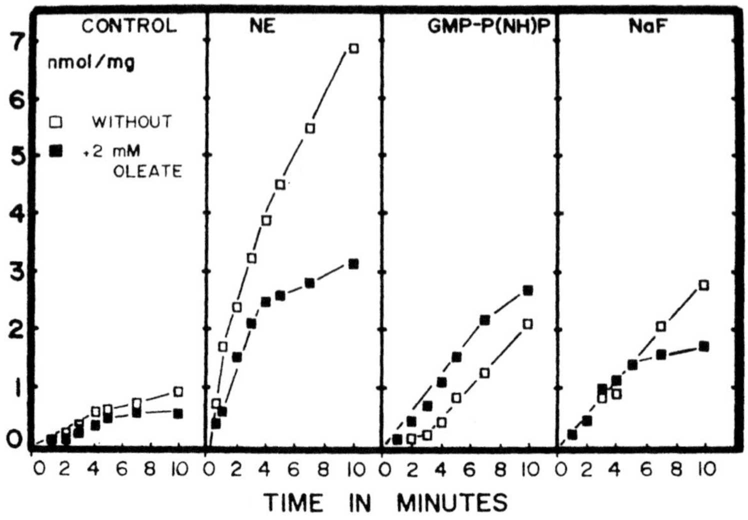


*Figure 10. Effect of oleate on activation of adenylate cyclase. Fat cell ghosts (46 μg/protein/tube) were incubated at 37°C without or with 200 μM norepinephrine, 10 μM GMP-P(NH)P, or 10 mM NaF in the absence (open squares) or presence (filled squares) of 2 mM oleate for the designated times. Ghosts were prepared and incubated as described in Figure 9. The values are expressed as nanomoles of cyclic AMP formed per mg protein and represent the means of three experiments.*

with the binding of catecholamines to their receptor or interferes with the activation of cyclase by the amine-receptor complex.

Further support for the hypothesis that free fatty acids can be important feedback regulators of adenylate cyclase came from studies with chicken fat cells (Malgieri *et al.*, 1975). In chicken fat cells, glucagon alone produced a large rise in cyclic AMP which was maintained over incubation periods of up to one hour. This suggested that chicken fat cells might not show the feedback regulation of cyclic AMP accumulation by lipolytic agents which is readily observed in rat fat cells.

The addition of enough sodium oleate to raise the free fatty acid to albumin ratio to 8 did not affect the level of cyclic AMP in chicken fat cells under conditions in which cyclic AMP accumulation by rat fat cells was markedly reduced (Malgieri *et al.*, 1975). Free fatty acid to albumin ratios as high as 12 did not inhibit adenylate cyclase activity of chicken fat cell ghosts (Malgieri *et al.*, 1975).

Our conclusion is that the prolonged elevation of cyclic AMP seen when chicken fat cells are incubated with lipolytic agents results from an absence of feedback regulation of chicken fat cell adenylate cyclase by the fatty acid/albumin ratio. The inhibition of lipolysis by elevated FFA/albumin ratios involves more than inhibition of cyclic AMP accumulation. Malgieri *et al.* (1975) found a profound inhibition of triglyceride lipase activity in rat fat cell homogenates by FFA/albumin ratios in the range of 3 to 6. This is illustrated in Figure 11 where more than 50% inhibition of lipase activity was seen with an FFA/albumin ratio of 3. There was only a small stimulation of lipase by the addition of cyclic AMP and ATP to the rat triglyceride lipase preparation and this was abolished by FFA/albumin ratios of 3 and 6.

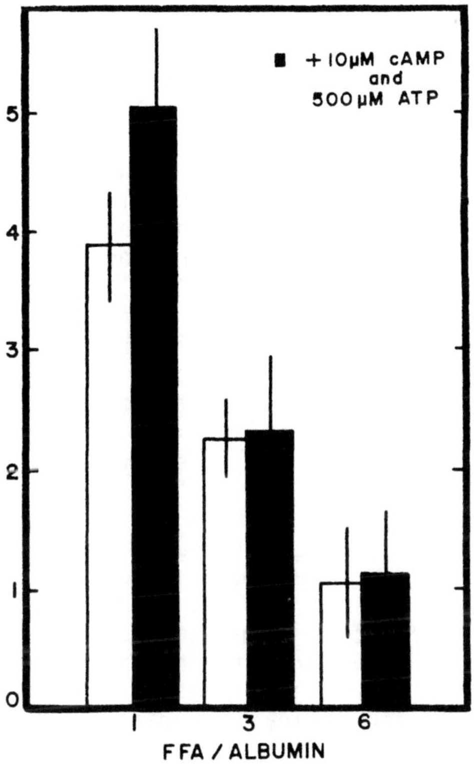


*Figure 11. Inhibition of triglyceride lipase by oleate. The pH 5.2 triglyceride lipase preparation (180 μg protein/tube) from rat fat cells was isolated by the procedure of Khoo and Steinberg (1974). The preparation was first incubated for 10 min at 30° in medium containing 5 mM $MgCl_2$, 1 mM theophylline, 1 mM dithiothreitol, 500 μM EGTA [ethylene glycol bis (β-amino-ethyl ether)-N, N-tetracetic acid], and 50 mM Tris-HCl, pH 8.0, without or with 10 μM cyclic AMP and 500 μM ATP. The assay was started by adding 0.6 ml of a substrate mixture containing 0.15 μmole of triolein emulsion (Carboxyl labeled [$^{14}C$]triolein in 5% gum arabic), 40 μmoles of sodium phosphate buffer (pH 6.8), and 20 mg of albumin either without or with added oleate to give a final FFA/albumin ratio of 1, 3, or 6. The tubes were incubated in a final volume of 0.8 ml for 30 min at 30°C. The final albumin concentration was 0.36 mM. The values are the means ± standard errors of five experiments. Hormone-sensitive lipase activity is expressed as micromoles FFA produced per mg protein over one hour in the absence (clear bars) or presence of 10 μM cyclic AMP and 500 μM ATP (filled bars).*

*Table 4: SUMMARY OF EFFECTS SEEN WITH OLEATE, OCTANOATE, AND LACTATE*

| Added acid | Lipolysis in fat cells | Cyclic AMP accumulation by fat cells | Adenylate cyclase activity fat cell ghosts | Triglyceride lipase activity |
|---|---|---|---|---|
| Oleate | inhibits | inhibits | markedly inhibits | inhibits |
| Octanoate | inhibits | inhibits | inhibits | no effect |
| Lactate | no effect | inhibits | no effect | no effect |

The effects of oleate (used as a representative long-chain fatty acid) are summarized in Table 4 and compared to those of octanoate (a short-chain fatty acid with eight carbons) and lactate. Fain and Shpherd (1976) reported that both lactate and octanoate mimicked the inhibitory effects of oleate on cyclic AMP accumulation by intact rat fat cells. There was no inhibition by either acid of cyclic AMP accumulation by chicken fat cells. Actually- there was an enhancement of the rise in cyclic AMP due to glucagon in chicken fat cells incubated with lactate. However, the mechanism by which lactate inhibits cyclic AMP accumulation was different from that for long-chain fatty acids since no direct inhibition of adenylate cyclase was observed in the presence of lactate (Fain and Shepherd, 1976).

Another striking difference between oleate and either octanoate or lactate was the failure of the latter two acids to inhibit the triglyceride lipase activity of fat cell homogenates (Fain and Shepherd, 1976). Of further interest was the finding that under our experimental conditions lipolysis was not inhibited by lactate.

A perplexing problem is the relationship of free fatty acids to the feedback regulator of adenylate cyclase originally described by Ho and Sutherland (1971). This feedback regulator also binds tightly to albumin and is nondialyzable. Ho, Bomboy, Wasner, and Sutherland (1975) reported that this feedback regulator could be largely separated from free fatty acids by partition chromatography on Sephadex LH-20 columns using a mobile phase of acetone-ethanol-heptane-water (4:50:1:9). This chromatographically purified feedback regulator was like fatty acids in that it gave opaque suspensions which were heat stable. However, the feedback regulator was acid-labile since 90% of the antagonist activity was lost upon exposure to pH1 for only 10 *sec* (Ho *et al.*, 1975).

One explanation is that the feedback regulator of Ho is an artifactual inhibitor formed from free fatty acids during its isolation and purification. This would account for the appearance of feedback regulator during the stimulation of fat cells by lipolytic agents including dibutyryl cyclic AMP (Ho *et al.*, 1975). Perhaps there was no appearance of feedback regulator in the absence of lipolytic agents since Ho *et al.* (1975) incubated fat cells in albumin freed of fatty acids by extraction with petroleum ether and acetic acid. In contrast, Fain and Shepherd (1975) found that a potent inhibitor of cyclic AMP accumulation could be obtained by the original procedure of Ho and Sutherland (1971) after incubation of fat cells without lipolytic agents using nondefatted albumin in the medium. In fact, the presence of fat cells was not necessary to obtain an inhibitor of cyclic AMP accumulation as it was sufficient to extensively dialyze albumin against distilled water and then lyophilize the albumin (Fain and Shepherd, 1975). Whether the inhibitor was formed from fatty acids present in the albumin is not known. What is clear is that artifical inhibitors of cyclic AMP accumulation can be readily formed by treatment of albumin containing fatty acids.

The feedback regulator of Ho *et al.* (1975) could be a physiologically important compound quite distinct from free fatty acids or adenosine. Since the feedback regulator is apparently some small molecular weight lipid-like substance, its relationship to prostaglandins is of interest. Ho and Sutherland (1971) and Ho *et al.* (1975) claim that the feedback regulator differs from prostaglandins. However, the feedback regulator could be one of the endoperoxides or thromboxanes formed from arachidonic acid rather than a prostaglandin of the E series. Gorman *et al.* (1975) have shown that, at fairly high concentrations endoperoxides formed from arachidonic acid are inhibitors of fat cell adenylate cyclase.

Our hypothesis is that the feedback regulator of Ho is formed by nonspecific peroxidation of polyunsaturated fatty acids released after activation of triglyceride lipase by hormones. Whether this compound(s) is formed by fat cells or by the procedures (10 *min* heating at 90°C and extensive dialysis against water) used by Ho for the extraction of the feedback regulator remains to be established.

## Cyclic GMP and the Anti-lipolytic Action of Insulin

Our view is that there is not yet any proof for a physiological role of cyclic AMP in insulin regulation of fat cell lipolysis. However, there is evidence that insulin can reduce cyclic AMP. Butcher, Sneyd, Park, and Sutherland (1966) originally reported that insulin could lower cyclic AMP in adipose tissue. Jungas (1966) shortly thereafter reported that prior treatment of adipose tissue with insulin reduced adenylate cyclase and glycogen phosphorylase while increasing glycogen synthetase in adipose tissue homogenates. It has also been demonstrated that the cyclic AMP phosphodiesterase activity of fat cell homogenates is enhanced by exposure to insulin prior to homogenization (Kono and Barham, 1973; Loten and Sneyd, 1970; Manganiello and Vaughan, 1973; Pawlson *et al.*, 1974; Sakai, Thompson, Lavis, and Williams, 1974; Solomon, 1975; Zinman and Hollenberg, 1974). However, no direct effects of insulin on the enzyme have been observed in cell free preparations. The effect of insulin is primarily on the phosphodiesterase activity of particles derived from the endoplasmic reticulum rather than the plasma membrane (Kono and Barham, 1973).

What relationship does the inhibition of adenylate cyclase and activation of cyclic AMP phosphodiesterase by insulin have to its physiological effects? While some reports have clearly shown an inhibition of net cyclic AMP accumulation by fat cells in the presence of insulin (Butcher *et al.*, 1966; Desai, Li, and Angel, 1973; Kono and Barham, 1973; Siddle and Hales, 1974) others have observed that under conditions where insulin was equally antilipolytic there was no decrease in cyclic AMP content (Fain and Rosenberg, 1972; Khoo *et al.*, 1973; Knight and Iliffe, 1973). Wieser and Fain (1975) compared the antilipolytic action of insulin to that of prostaglandin $E_1$, nicotinic acid, and adenosine. While all four agents were antilipolytic there was no inhibition of cyclic AMP accumulation by insulin whereas the other agents markedly reduced cyclic AMP. The available evidence indicates that an antilipolytic action of insulin can be seen under conditions in which total cyclic AMP accumulation is unaltered. There is some interaction between insulin and cyclic AMP in that the stimulation of glucose metabolism and inhibition of lipolysis by insulin is reduced under conditions in which cyclic AMP is elevated (Fain and Wieser, 1975; Wieser and Fain, 1975).

Fain (1975) found that incubation of fat cells with insulin and lipolytic agents actually enhances the ability of fat cells to elevate cyclic AMP accumulation after a second addition of adenylate cyclase activators. If fat cells were incubated with lipolytic agents under conditions in which the final FFA/albumin ratio exceeded three, the rise in cyclic AMP accumulation due to a second hormone addition was markedly reduced. If insulin was also present the FFA/albumin ratio remained below three and there was no inhibition of the rise in cyclic AMP accumulation due to a second addition of lipolytic hormones. Incubation of fat cells with prostaglandin $E_1$ had quite different effects as it inhibited both lipolysis and the rise in cyclic AMP accumulation due to a second addition of hormones (Fain, 1975).

Illiano, Tell, Siegel, and Cuatrecasas (1973) reported that

insulin and carbachol (a long lasting cholinergic agent) elevated cyclic GMP in rat fat cells. Fain and Butcher (1976) and Vydelingum, Kissebah, Wynn, and Simpson (1975) found that insulin increased cyclic GMP in fat cells. Fain and Butcher (1976) confirmed that carbachol elevated cyclic GMP. Furthermore, cyclic GMP was also elevated in fat cells by the divalent cation ionophore A23187, norepinephrine, and oleate (Fain and Butcher, 1976).

The problem in invoking any role of cyclic GMP in the regulation of lipolysis is the lack of correlation between cyclic GMP and lipolysis. Chlouverakis (1963) first reported that cholinergic agents did not affect fat cell lipolysis. Subsequently, this was confirmed by Fain and Butcher (1976) who also found that A23187 had no inhibitory effect on lipolysis. There is no apparent correlation between lipolysis and cyclic GMP. Cyclic GMP is increased by agents which either activate, inhibit, or have no effect on lipolysis (Fain and Butcher, 1976).

Increases in cyclic GMP accumulation due to norepinephrine, carbachol, insulin, oleate, and A23187 were markedly inhibited by incubation of fat cells in the calcium chelator EGTA (Fain and Butcher, 1976). However, lipolysis due to catecholamines was unaffected by incubation in $Ca^{2+}$-free buffer containing EGTA (Fain and Butcher, 1976; Khoo *et al.*, 1973).

Vydelingum *et al.* (1975) observed that the increase in cyclic GMP due to insulin was abolished in $Ca^{2+}$-free buffer as was that of leucine incorporation into protein. Their results suggested that insulin-induced protein synthesis might require the presence of extracellular $Ca^{2+}$ and invoke cyclic GMP. However, the antilipolytic action of insulin was not affected by omission of extracellular $Ca^{2+}$ (Fain and Butcher, 1976).

Our view of insulin action is summarized in Figure 12. Insulin probably binds to specific receptors in the plasma membrane and the insulin-receptor complex alters the activity of a wide number of membrane bound enzymes. There may be no unique second messenger for insulin since all the effects could be the result of propagated membrane disturbances. The inhibitory effect of insulin on lipolysis is shown as a direct effect on triglyceride lipase which is unrelated to the observed elevation in cyclic GMP or the rise in cyclic AMP phosphodiesterase activity due to insulin.

## Summary

The regulation of lipolysis in the fat cells of adipose tissue appears to be the key step in the utilization of stored energy during starvation and other conditions. Lipolysis is directly activated by catecholamines and the sensitivity to catecholamines re-

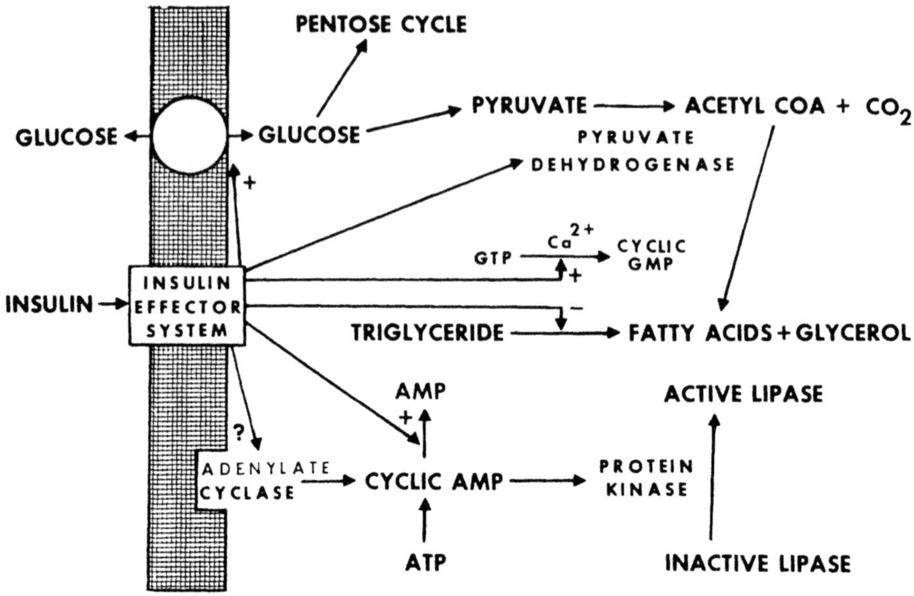


*Figure 12. Insulin action on fat cell metabolism. The figure illustrates six possible effects of insulin on fat cell metabolism and suggests that all are the result of a primary effect on membrane-bound enzymes. It is quite possible that some of the effects are more interrelated than is depicted in the model. The six effects of insulin listed are (listed in clockwise fashion): (1) stimulation of glucose transport, (2) activation of pyruvate dehydrogenase, (3) increase in cyclic GMP via increase in intracellular* $Ca^{2+}$*, (4) inhibition of triglyceride lipase, (5) activation of cyclic AMP phosphodiesterase, and (6) inhibition of adenylate cyclase.*

leased at nerve endings is modulated by thyroid hormones, glucocorticoids, growth hormone, and insulin. Insulin inhibits adipose tissue fatty acid release both by inhibiting lipolysis and accelerating fatty acid re-esterification.

The known mechanism for activation of lipolysis by catecholamines involves activation of adenylate cyclase followed by an increase in protein kinase which phosphorylates inactive triglyceride lipase to an active lipase. There appear to be other factors involved in catecholamine activation of lipolysis in isolated rat fat cells since detectable elevations of cyclic AMP are seen only in the presence of methylxanthines. Lipolysis can be maximally activated by catecholamines without any rise in cyclic AMP. Furthermore, the rise in cyclic AMP due to catecholamines in the presence of methylxanthines can be blocked by adenosine without lipolysis being inhibited.

Adenosine is unlikely to be a feedback regulator of adenylate cyclase released during incubation of fat cells with lipolytic hormones. We observe an effect of hormones on adenosine accumulation by rat fat cells only when AMP is elevated and adenosine deamination to inosine is blocked. Chicken fat cells appear to release less adenosine and deaminate more adenosine to inosine than do rat fat cells. The finding that chicken fat cells have little adenosine present in the incubation medium probably accounts for the lack of a response to adenosine deaminase and theophylline.

If analogs of adenosine that cannot be deaminated are added to incubated chicken fat cells, or if adenosine is added in sufficient amounts, the increase in cyclic AMP accumulation due to glucagon is inhibited and theophylline is able to reverse the adenosine effect. These data suggest that one site for theophylline potentiation of cyclic AMP accumulation could be an antagonism of adenosine action. We postulate that endogenous adenosine partially inhibits adenylate cyclase and methylxanthines elevate cyclic AMP by blocking adenosine action in rat fat cells.

The inhibitory effect of adenosine on cyclic AMP accumulation is apparently unrelated to adenosine uptake into fat cells or conversion to nucleotides. Adenosine can be bound to stachyose and the resulting large water-soluble compound is fully as active as adenosine in inhibiting cyclic AMP accumulation in rat or chicken fat cells.

The rise in cyclic AMP in rat fat cells due to catecholamines in the presence of methylxanthines is transitory and cells are refractory to further stimulation of cyclic AMP by re-addition of hormones. Fat cells release inhibitors of cyclic AMP accumulation to the medium during incubation with lipolytic agents. Our data suggest that these feedback regulators are free fatty acids and when the FFA/albumin ratio exceeds 3-5 there is an inhibition of both adenylate cyclase and triglyceride lipase.

Chicken fat cells do not show a rise followed by a sharp drop in cyclic AMP accumulation. Furthermore, the incubation of chicken fat cells in medium with an FFA/albumin ratio of 5-8 does not inhibit lipolysis or cyclic AMP accumulation while both are inhibited in similar experiments with rat fat cells. The adenylate cyclase activity of chicken fat cell ghosts incubated with glucagon was not inhibited by FFA/albumin ratios of 4-8 while the cyclase activity of rat fat cell ghosts was markedly inhibited. We attribute the failure of chicken fat cells to demonstrate feedback regulation of cyclic AMP accumulation to an insensitivity of adenylate cyclase to inhibition by free fatty acids.

The activation of adenylate cyclase in fat cell ghosts and plasma membrane preparations by GMP-P(NH)P was actually enhanced by oleate. The four-minute lag period in activation of adenylate cyclase by GMP-P(NH)P was reduced by oleate under conditions in which the activation of cyclase by norepinephrine was markedly inhibited. These data suggest that oleate does not inhibit the catalytic activity of adenylate cyclase but rather the binding or interaction of catecholamines with the enzyme.

Our conclusion is that the major feedback regulator of adenylate cyclase is free fatty acids released during lipolysis. Adenosine does not appear to play a key role in feedback regulation but the large elevation of cyclic AMP seen in the presence of methylxanthines may result from blockade of adenosine inhibition of adenylate cyclase.

The mechanism by which insulin inhibits lipolysis remains to be established. The effects of insulin are quite different from those of other antilipolytic agents such as adenosine, prostaglandins, and nicotinic acid. While it is established that prior treatment of fat cells with insulin can activate cyclic AMP phosphodiesterase and lower cyclic AMP, there is no clear proof that these effects explain the antilipolytic action on insulin. Under other conditions insulin inhibits lipolysis without affecting cyclic AMP accumulation. Insulin does elevate cyclic AMP in fat cells but so does carbachol which has no antilipolytic effect. We favor the hypothesis that the increases in cyclic GMP and decreases in cyclic AMP due to insulin addition to fat cells are unrelated to the mechanism for the antilipolytic action of insulin.

## REFERENCES

Angel, A., Desai, K.S., and Halperin, M.L. (1971). Reduction in adipocyte ATP by lipolytic agents: Relation to intracellular free fatty acid accumulation. *J. Lipid Res.* 12:203-211.

Birnbaumer, L. and Rodbell, M. (1969). Adenyl cyclase in fat cells. II. Hormone receptors. *J. Biol. Chem.* 244:3477-3482.

Boyd, T.A., Weiser, P.B., and Fain, J.N. (1975). Lipolysis and cyclic AMP accumulation in isolated fat cells from chickens. *Gen. Comp. Endocrinol.* 26:243-247.

Burns, T.W., Langley, P.E., and Robison, G.A. (1975). Site of free-fatty acid inhibition of lipolysis by human adipocytes. *Metabolism* 24:265-276.

Butcher, R.W. and Baird, C.E. (1969). The regulation of cyclic AMP and lipolysis in adipose tissue by hormones and other agents. pp. 5-23. In W.L. Holmes, L.A. Carlson, and R. Paoletti (Eds.) *Drugs Affecting Lipid Metabolism.* Plenum Press, New York.

Butcher, R.W., Sneyd, J.G.T., Park, C.R., and Sutherland, E.J. Jr. (1966). Effect of insulin on adenosine 3'5'-monophosphate in the rat epididymal fat pad. *J. Biol. Chem.* 241:1651-1653.

Chlouverakis, C. (1963). Parasympathomimetic agents and the metabolism of rat adipose tissue. *Metabolism* 12:936-940.

Corbin, J.D., Reimann, E.M., Walsh, D.A., and Krebs, E.G. (1969). A cyclic AMP-stimulated protein kinase in adipose tissue. *Biochem. Biophys. Res. Commun.* 36:328-336.

Dalton, C. and Hope, H.R. (1973). Inability of prostaglandin synthesis inhibitors to affect adipose tissue lipolysis. *Prostaglandins* 4:641-651.

Dalton, C. and Hope, W.C. (1974). Cyclic AMP regulation of prostaglandin biosynthesis in fat cells. *Prostaglandins* 6:227-242.

Desai, K.S., Li, K.C., and Angel, A. (1973). Bimodal effect of insulin on hormone-stimulated lipolysis: Relation to intracellular 3',5'-cyclic adenylic acid and free fatty acids levels. *J. Lipid Res.* 14:674-655.

Dole, V.P. (1956). A relation between non-esterified fatty acids in plasma and the metabolism of glucose. *J. Clin. Invest.* 35:150-154.

Fain, J.N. (1971). Effects of menadione and vitamin $K_5$ on glucose metabolism, respiration, lipolysis, cyclic 3',5'-adenylic acid accumulaiton, and adenyl cyclase in white fat cells. *Mol. Pharmacol.* 7:465-479.

Fain, J.N. (1973a). Biochemical aspects of drug and hormone action on adipose tissue. *Pharmacol. Rev.* 25:67-118.

Fain, J.N. (1973b). Inhibition of adenosine cyclic 3'5'-monophosphate accumulation in fat cells by adenosine, $N^6$ (phenylisopropyl) adenosine, and related compounds. *Mol. Pharmacol.* 9:595-604.

Fain, J.N. (1975). Insulin as an activator of cyclic AMP accumulation in rat fat cells. *J. Cyclic Nuc. Res.* 1:359-366.

Fain, J.N. (1977). Cyclic nucleotides in adipose tissue. In H. Cramer and J. Schultz (Eds.) *Cyclic Nucleotides: Mechanisms of Action.* John Wiley, London.

Fain, J.N. and Butcher, F.R. (1976). Cyclic guanosine 3',5'-monophosphate and the regulation of lipolysis in rat fat cells. *J. Cyclic Nuc. Res.* 2:71-78.

Fain, J.N., Dodd, A., and Novak, L. (1971). Enzyme regulation in gluconeogenesis and lipogenesis: Relationship of protein synthesis and cyclic AMP to lipolytic growth hormone and glucocorticoids. *Metabolism* 20:109-118.

Fain, J.N. and Loken, S.C. (1971). Valinomycin as an antilipolytic agent. Effects on cyclic 3',5'-adenylic acid, adenyl cyclase, respiration,and glucose metabolism in brown and white fat cells. *Mol. Pharmacol.* 7:455-464.

Fain, J.N., Pointer, R.H., and Ward, W.F. (1972). Effects of adenosine nucleotides on adenylate cyclase, phosphodiesterase, cyclic adenosine monophosphate accumulation, and lipolysis in fat cells. *J. Biol. Chem.* 247:6866-6872.

Fain, J.N., Psychoyos, S., Czernik, A.J., Frost, S., and Cash, W.D. (1973). Indomethacin, lipolysis and cyclic AMP accumulation in white fat cells. *Endocrinology* 93:632-639.

Fain, J.N. and Rosenberg, L. (1972). Antilipolytic action of insulin on fat cells. *Diabetes* 21:414-425.

Fain, J.N. and Saperstein, R. (1970). The involvement of RNA synthesis and cyclic AMP in the activation of fat cell lipolysis by growth hormone and glucocorticoids. pp. 20-27. In B. Jeanrenaud and D. Hepp (Eds.) *Adipose Tissue: Regulation and Metabolic Functions*. Academic Press, New York.

Fain, J.N. and Shepherd, R.E. (1975). Free fatty acids as feedback regulators of adenylate cyclase and cyclic 3',5'-AMP accumulation in rat fat cells. *J. Biol. Chem.* 250:6586-6592.

Fain, J.N. and Shepherd, R.E. (1976). Inhibition of adenosine 3',5'-monophosphate accumulation in white fat cells by short chain fatty acids, lactate and β-hydroxybutyrate. *J. Lipid Res.* 17:377-385.

Fain, J.N. and Wieser, P.B. (1975). Effects of adenosine deaminase on cyclic adenosine monophosphate accumulation, lipolysis and glucose metabolism of fat cells. *J. Biol. Chem.* 250: 1027-1034.

Fredholm, B.B. and Hedquist, P. (1975). Indomethacin and the role of prostaglandins in adipose tissue. *Biochem. Pharmacol.* 24:61-66.

Gordon, R.S. Jr. and Cherkes, A. (1956). Unesterified fatty acids in human blood plasma. *J. Clin. Invest.* 35:206-212.

Gorman, R.R. (1975). Prostaglandin endoperoxides: Possible new regulators of cyclic nucleotide metabolism. *J. Cyclic Nuc. Res.* 1:1-9.

Gorman, R.R., Hamberg, M., and Samuelson, B. (1975). Inhibition of basal and hormone-stimulated adenylate cyclase in adipocyte ghosts by the prostaglandin endoperoxide prostaglandin $H_2$. *J. Biol. Chem.* 250:6460-6463.

Hamberg, M. and Samuelsson, B. (1974). Prostaglandin endoperoxides. Novel transformation of arachidonic acid in human platelets. *Proc. Nat. Acad. Sci.* U.S.A. 71:3400-3404.

Haslam, R.J. and Rosson, G.M. (1975). Effects of adenosine on levels of adenosine cyclic 3',5'-monophosphate in human blood platelets in relation to adenosine incorporation and platelet aggregation. *Mol. Pharm.* 11:528-544.

Ho, R.J., Bomboy, J.D., Wasner, H.K., and Sutherland, E.W. (1975). Preparation and characterization of a hormone antagonist from adipocytes. *Meth. in Enzymol.* 39:431-438.

Ho, R.J. and Sutherland, E.W. (1971). Formation and release of a hormone antagonist by rat adipocytes. *J. Biol. Chem.* 246: 6822-6827.

Illiano, G. and Cuatrecasas, P. (1971). Endogenous prostaglandins modulate lipolytic processes in adipose tissue. *Nature New Biol.* 234:72-74.

Illiano, G., Tell, G.P.E., Siegel, M.I., and Cuatrecasas, P. (1973). Guanosine 3',5'-cyclic monophosphate and the action of insulin. *Proc. Nat. Acad. Sci.* U.S.A. **70**:2443-2447.

Jungas, R.L. (1966). Role of cyclic-3'5'-AMP in the response of adipose tissue to insulin. *Proc. Nat. Acad. Sci.* U.S.A. **56**: 757-763.

Khoo, J.C., Aguino, A.A., and Steinberg, D. (1974). The mechanism of activation of hormone-sensitive lipase in human adipose tissue. *J. Clin. Invest.* **53**:1124-1131.

Khoo, J.C. and Steinberg, D. (1974). Reversible protein kinase activation of hormone-sensitive lipase from chicken adipose tissue. *J. Lipid Res.* **15**:602-610.

Khoo, J.C., Steinberg, D., Thompson, B., and Mayer, S.E. (1973). Hormonal regulation of adipocyte enzymes. *J. Biol. Chem.* **248**:3823-3830.

Kitabgi, P., Rosselin, G., and Bataille, D. (1976). Interactions of glucagon and related peptides with chicken adipose tissue. *Horm. Metab. Res.* **8**:266-270.

Knight, B.L. and Iliffe, J. (1973). The effect of glucose, insulin and noradrenaline on lipolysis and on the concentrations of adenosine 3',5'-monophosphate and adenosine 5'-triphosphate in adipose tissue. *Biochem. J.* **132**:77-82.

Kono, T. and Barham, F.W. (1973). Effects of insulin on the levels of adenosine 3'5'-monophosphate and lipolysis in isolated rat epididymal fat cells. *J. Biol. Chem.* 248:7417-7426.

Lamberts, S.W.J., Timmermans, H.A.T., Kramer-Blankestijn, M., and Birkenhager, J. (1975). The mechanism of the potentiating effect of glucocorticoids on catecholamine-induced lipolysis. *Metabolism* **24**:681-689.

Loten, E.G. and Sneyd, J.G.T. (1970). An effect of insulin on adipose-tissue adenosine 3'5-cyclic monophosphate phosphodiesterase. *Biochem. J.* **120**:187-193.

Malgieri, J.A., Shepherd, R.E., and Fain, J.N. (1975). Lack of feedback regulation of cyclic 3'5'AMP accumulation by free fatty acids in chicken fat cells. *J. Biol. Chem.* **250**:6593-6598.

Manganiello, V., Murad, F., and Vaughan, M. (1971). Effects of lipolytic and antilipolytic agents on cyclic 3'5'-adenosine. *J. Biol. Chem.* **246**:2195-2202.

Manganiello, V. and Vaughan, M. (1973). An effect of insulin on cyclic adenosine 3'5'-monophosphate phosphodiesterase activity in fat cells. *J. Biol. Chem.* **248**:7164-7170.

Moskowitz, J. and Fain, J.N. (1970). Stimulation by growth hormone and dexamethasone of labeled cyclic adenosine 3'5'-monophosphate accumulation by white fat cells. *J. Biol. Chem.* **245**:1101-1107.

Namm, D.H. and Leader, J.P. (1974). A sensitive analytical method for the detection and quantitation of adenosine in biological samples. *Anal. Biochem.* **58**:511-524.

Olsson, R.A., Davis, C.J., Khouri, E.M., and Patterson, R.E. (1976). Evidence for an adenosine receptor on the surface of dog coronary myocytes. *Circulat. Res.* 39:93-98.

Paterson, A.R.P. and Oliver, J.M. (1971). Nucleoside transport. II. Inhibition by p-nitrobenzyl-thioguanosine and related compounds. *Can. J. Biochem.* 49:271-274.

Pawlson, L.G., Lovell-Smith, C.J., Manganiello, V.C.,and Vaughan, M. (1974). Effects of epinephrine, adrenocorticotrophic hormone, and theophylline on adenosine 3'5'-monophosphate phosphodiesterase activity in fat cells. *Proc. Nat. Acad. Sci.* U.S.A. 71:1639-1642.

Rodbell, M. (1965). Modulation of lipolysis in adipose tissue by fatty acid concentration in fat cell. *Ann. N.Y. Acad. Sci.* 131:302-333.

Rodbell, M. (1967). Metabolism of isolated fat cells. V. Preparation of "ghosts" and their properties; adenyl cyclase and other enzymes. *J. Biol. Chem.* 242:5744-5750.

Rodbell, M. (1975). On the mechanism of activation of fat cell adenylate cyclase by guanine nucleotides. *J. Biol. Chem.* 250:5826-5834.

Sakai, T., Thompson, W.J., Lavis, V.R., and Williams, R.H. (1974). Cyclic nucleotide phosphodiesterase activities from isolated fat cells: Correlation of subcellular distribution with effects of nucleotides and insulin. *Arch. Biochem. Biophys.* 162:331-339.

Schaeffer, H.J. and Schwender, C.F. (1974). Enzyme inhibitors, 26, bridging hydrophobic and hydrophilic regions on adenosine deaminase with some 9-(2-hydroxy-3-alkyl) adenines. *J. Med. Chem.* 17:6-8.

Schwabe, U., Ebert, R., and Erbler, H.C. (1973). Adenosine release from isolated fat cells and its significance for the effects of hormones on cyclic 3'5-AMP levels and lipolysis. *Arch. Pharmacol.* 276:133-148.

Siddle, K. and Hales, C.N. (1974). The relationship between the concentration of adenosine 3'5'-cyclic monophosphate and the anti-lipolytic action of insulin in isolated rat fat-cells. *Biochem. J.* 142:97-103.

Solomon, S.S. (1975). Effect of insulin and lipolytic hormones on cyclic AMP phosphodiesterase activity in normal and diabetic rat adipose tissue. *Endocrinology* 96:1366-1373.

Stock, K. and Prilop, M. (1974). Dissociation of catecholamine-induced formation of adenosine 3'5'-monophosphate and release of glycerol in fat cells by prostaglandin $E_1$, $E_2$ and $N^6$ phenylisopropyladenosine. *Arch. Pharmacol.* 282:15-31.

Vydelingum, N., Kissebah, A.H., Wynn, V., and Simpson, A. (1975). The role of calcium in insulin action. V. Insulin, calcium c-GMP and the regulation of protein synthesis in adipose tissue. *Diabetologia* 11:382 (Abstract).

Wieser, P.B. and Fain, J.N. (1975). Insulin, prostaglandin $E_1$, phenylisopropyladenosine and nicotinic acid as regulators of fat cell metabolism. *Endocrinology* 96:1221-1225.

Zinman, B. and Hollenberg, C.H. (1974). Effect of insulin and lipolytic agents on rat adipocyte low $K_m$ cyclic adenosine 3'5',-monophosphate phosphodiesterase. *J. Biol. Chem.* 249: 2182-2187.

# HORMONAL CONTROL OF KETOGENESIS

J. Denis McGarry and Daniel W. Foster

Departments of Internal Medicine and Biochemistry
University of Texas Health Science Center
5323 Harry Hines Blvd.; Dallas, TX 75235

## OUTLINE



## I. INTRODUCTION

Interest in how the body regulates its production of the ketone bodies, acetoacetic and β-hydroxybutyric acids, has grown exponentially in recent years both among clinicians and biochemists. The reason stems from the fact that historically these compounds have been associated primarily with pathological disorders, most notably uncontrolled diabetes, where their excessive accumulation in the blood constitutes a life-threatening problem. During the past two decades, however, it has been clearly established that in certain physiological situations a modest elevation in the blood ketone level is crucial to the overall fuel economy of the body

(Bailey and Lockwood, 1973; Krebs, 1966; Krebs, Williamson, Bates, Page, and Hawkins, 1971; Owen, Felig, Morgan, Wahren, and Cahill, 1969; Owen, Morgan, Kemp, Sullivan, Herrera, and Cahill, 1967). Nevertheless, despite these major advances a number of fundamental issues concerning the hormonal and biochemical mechanisms involved in the control of the ketogenic process remain to be clarified. In the present paper we shall focus on recent developments that promise to shed new light on this area of metabolic regulation.

## II. GENERAL REQUIREMENTS FOR MAXIMAL KETOGENESIS

It has long been known that the ketotic state is characterized by a partial or absolute deficiency of insulin. Accordingly, a widely held view of the events leading up to the development of ketosis has been that depicted in Figure 1. There is little doubt as to the validity of the initial steps in this formulation. Thus, since ketone bodies arise almost exclusively from the oxidation of fatty acids in liver, it is evident that elevation of the plasma free fatty acid concentration is a *sine qua non* for rapid rates of ketogenesis to occur. Nor is it disputed that accelerated mobilization of free fatty acids from adipose tissue requires a fall in the circulating level of insulin. A misleading feature of the scheme, however, and one that is still not generally recognized, is the implication that the output of ketone bodies by the liver is governed simply by the input of fatty acids. Convincing evidence against this notion was first provided by the studies of Lossow and Chaikoff (1954) and Lossow, Brown, and Chaikoff (1956) who demonstrated both *in vivo* and *in vitro* that, when exposed to the same concentration of long chain fatty acids, livers from ketotic rats oxidized a much greater fraction to ketone bodies than did those from normal, fed animals. Subsequently, Mayes and

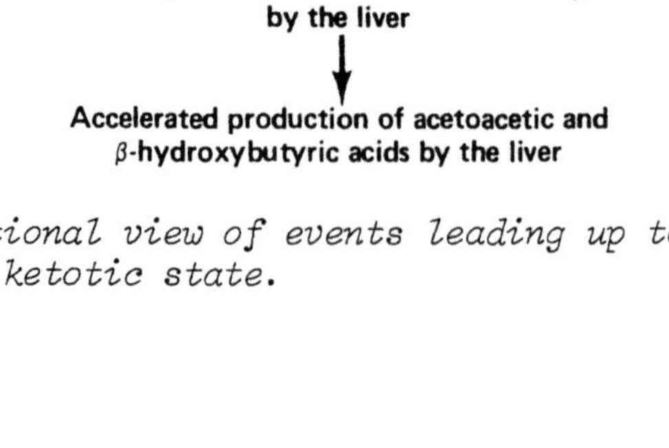


*Figure 1. Conventional view of events leading up to the development of the ketotic state.*

Felts (1967) showed that the different pattern of fatty acid metabolism in livers from normal and ketotic rats stemmed from a shift in the partitioning of the substrate between the pathways of esterification and oxidation, the latter being greatly favored in ketotic animals. The phenomenon has since been extensively confirmed by other groups (McGarry and Foster, 1971b; Ontko, 1972; Woodside and Heimberg, 1972). Clearly, then, high rates of ketone production in the whole organism require, in addition to adequate substrate availability, a dramatic change in hepatic metabolism such that the fatty acids taken up are efficiently converted into acetoacetic and β-hydroxybutyric acids. While a full understanding of the hormonal and biochemical basis for this altered metabolic profile of the "ketotic" liver has so far defied intensive investigation, some clues to the nature of the control processes have begun to emerge.

## III. HORMONAL FACTORS IN THE REGULATION OF KETOGENESIS

### A. The Roles of Insulin and Glucagon

Since activation of adipose tissue lipolysis and hepatic fatty acid oxidation capacity appear to occur synchronously whenever the plasma insulin level falls, we wanted to know whether both events are the direct result of insulin deficiency *per se*, or whether the antagonistic hormone, glucagon, might also be involved. To this end, fed, nonketotic rats were treated acutely with insulin antibodies or glucagon to determine the effects of these agents on a variety of metabolic parameters related to carbohydrate and lipid metabolism and on hepatic ketogenic capacity (defined as the ability of the liver to synthesize ketone bodies when perfused with oleic acid). Consistent with previous reports (Armin, Grant, and Wright, 1960; Balasse, Bier, and Havel, 1972; Williamson, 1967; Woodside and Heimberg, 1972) infusion of anti-insulin serum resulted in striking elevations of the plasma glucose, free fatty acid and ketone body concentrations, and marked depletion of liver glycogen stores (Figures 2A-D). The picture after glucagon treatment, however, was quite different in several respects. In this case there was only a small rise in the plasma glucose concentration (despite almost total depletion of liver glycogen) and essentially no effect on plasma free fatty acid and ketone levels. These results were not surprising when earlier studies were considered and can be explained as follows. Glucagon administration to fed animals triggers hepatic glycogenolysis and causes a transient elevation of the blood glucose concentration. This, in turn, elicits insulin release from the pancreatic β-cell which allows efficient removal of glucose from the blood in nonhepatic tissues and, in addition, inhibits lipolysis in fat depots, so that the liver is denied an increased supply of ketogenic substrate (Liljenquist, Bomboy, Lewis, Sinclair-Smith, Felts, Lacy, Crofford, and Liddle, 1974; Whitty,

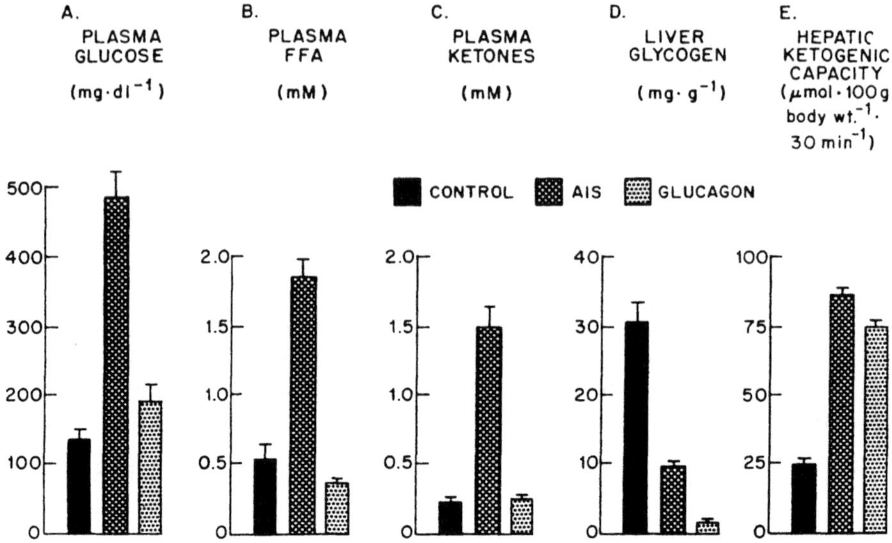


*Figure 2. Effects of anti-insulin serum (AIS) and glucagon infusion into fed rats. Animals (100-200 g) received intravenously guinea pig serum containing approximately 2 U of insulin antibody per mℓ, 100 μg of glucagon per mℓ or neither agent at a rate of 200 μℓ per minute for 5 minutes followed by 10 μℓ per minute for a period of 3 hours. At this point arterial blood was collected for analyses and livers were taken for measurement of their glycogen content or were perfused with oleic acid to determine their ketogenic capacity. Values are means ± SEM for 4-8 animals in each group. For full details see McGarry, Wright, and Foster (1975).*

Shima, Turbow, and Foà, 1969). The unique finding of our experiments, however, was that despite their markedly different effects *in vivo* both anti-insulin serum and glucagon caused a major (three- to four-fold) increase in hepatic ketogenic capacity (Figure 2E). This increase in ketogenic capacity could be induced by as little as one hour of treatment, a time that can be put in perspective by the observation that six to nine hours of fasting are required to induce similar changes without hormone treatment (McGarry, Meier, and Foster, 1973). Similar effects of anti-insulin serum in fed rats have recently been reported by Woodside and Heimberg (1976). That the absence of ketonemia in the glucagon-treated rats was due solely to hepatic insufficiency of substrate was established by infusion of intralipid plus heparin which raised the plasma free fatty acid concentration acutely to the region of 2-3 mM. Ketonemia immediately developed in those animals previously treated with glucagon while control animals showed a much more sluggish response (data not shown).

The above experiments demonstrated for the first time that a "ketogenic" liver could be obtained from a nonketotic animal. This unusual situation can be brought about under circumstances where circulating insulin levels are normal, or even elevated, provided and excess of glucagon is also available. The possibility thus exists that activation of the liver's ketogenic machinery in ketotic states is due, in large part, to the increase in circulating glucagon levels that invariably accompanies such conditions. This led us to propose that, contrary to previous beliefs, the ketotic state is not the result of a single hormone disturbance, namely insulin lack, but also requires for its development an inappropriate elevation in the blood glucagon concentration. This bihormonal concept for the control of ketone body production is summarized in Figure 3. Certain reservations should be kept in mind in the interpretation of this model. First, although only two hormones are emphasized, it is recognized that others, particularly those of pituitary and adrenal origin, are also important in the etiology of the ketotic state (Engel and Amatruda, 1963; Houssay, 1963; Scow and Chernick, 1960). At present, however, we believe it is unlikely that these act in the acute "on-off" control of the ketogenic process. Second, while control site 2 is depicted as being governed primarily by the circulating glucagon concentration, the possibility that insulin exerts an anti-ketogenic effect at the level of the liver is not excluded. Indeed, the functional unit here might well be the [glucagon]:[insulin] ratio. Despite these limitations, the basic features of the model have received strong support from studies in humans. For example, blockade of pancreatic

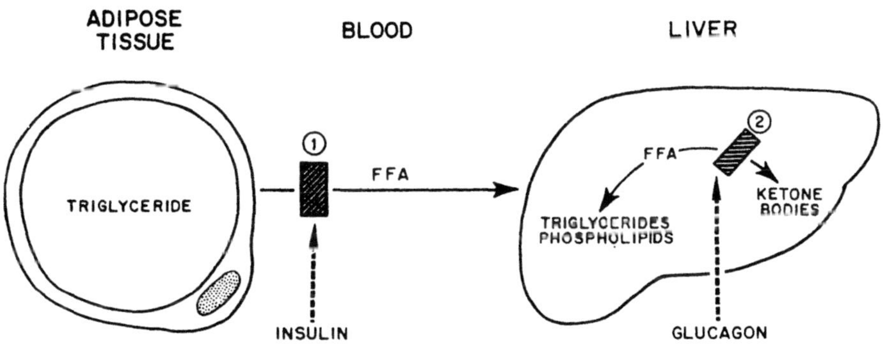


*Figure 3. Bihormonal model for the control of ketogenesis. Adipose tissue lipolysis (site 1) is considered to be controlled primarily by insulin (inhibition), while the ability of the liver to oxidize fatty acids to ketone bodies (site 2) is thought to be controlled, in large part, by glucagon (activation). Thus, elevation of the [glucagon] : [insulin] ratio provides the necessary conditions for accelerated ketogenesis.*

glucagon secretion with somatostatin has been shown to markedly obtund the development of hyperglycemia and ketosis after withdrawal of insulin from insulin-dependent diabetics (Gerich, Lorenzi, Bier, Schneider, Tsalikian, Karam, and Forsham, 1975). Conversely, administration of glucagon has been shown to enhance the ketogenic response to a fatty acid load in diabetic man (Schade and Eaton, 1975).

## B. Physiological *versus* Pathological Ketosis

Implicit in the scheme outlined in Figure 3 is that once the ketogenic machinery of the liver has been turned on (site 2), the production of ketone bodies will be determined by the extent of adipose tissue lipolysis (site 1) which, in turn, will be governed principally by the degree of insulin deficiency. Evidence to this effect is seen in clinical medicine. For example, during prolonged starvation in man the plasma-free fatty acid concentration rises to approximately 1 *mM* and remains at this level for several weeks. Concomitantly, the blood ketone level rises and plateaus at about 5-6 *mM*. The maintenance of this physiological steady state is apparently achieved through the operation of a negative feedback loop of the ketone bodies on the lipolytic process as depicted in Figure 4. To what extent acetoacetate and β-hydroxybutyrate act directly on adipose tissue or indirectly by stimulating pancreatic insulin secretion is not clear (Björntorp and Scherstén, 1967; Madison, Mebane, Unger, and Lochner, 1964). In severe ketoacidosis, on the

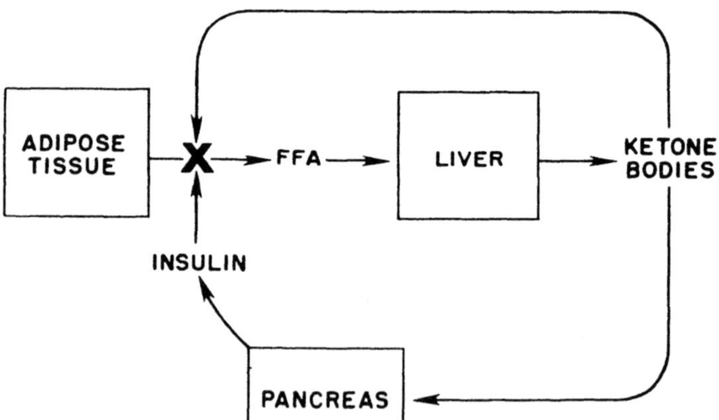


*Figure 4. Model for physiological ketosis. The ketone bodies modulate their own production through a negative feedback inhibition, directly or indirectly via insulin, on adipose tissue lipolysis.*

other hand, the feedback mechanism breaks down either because insulin production is defective (diabetes) or because other factors, such as epinephrine, block its release, simultaneously stimulating lipolysis (the syndrome of alcoholic ketoacidosis might fall into the latter category). The end result is a severe metabolic acidosis characterized by plasma free fatty acid and ketone body concentrations in the region of 2-3 *mM* and 15-20 *mM*, respectively Alberti, Hockaday, and Turner, 1973; Levy, Duga, Gingis, and Gordon, 1973; Meier, McGarry, and Foster, 1972). Although not yet proved, a similar sequence of events will undoubtedly be established in cases of severe ketosis during the late stages of pregnancy.

## IV. INTRAHEPATIC FACTORS IN THE REGULATION OF KETOGENESIS

### A. Esterification *versus* Oxidation--The Role of Carnitine Acyltransferase

From the above discussion it is obvious that an understanding of the biochemical switch responsible for "turning on" the ketogenic machinery of the liver requires elucidation of the factors that govern the channelling of fatty acids between the two major pathways open to them, namely, esterification in the extramitochondrial compartment and oxidation within the mitochondrion (Figure 5). The central question here is whether the increased ketogenic capacity of the ketotic liver (illustrated in Figure 6A and B) is a consequence of a diminished ability of the tissue to esterify incoming fatty acids, as originally suggested (Mayes and Felts, 1967; Tzur, Tal, and Shapiro, 1964), or the result of direct activation of the oxidative sequence. While it has not been possible to prove

*Figure 5. Major pathways for the metabolism of long-chain fatty acids in liver. Reactions 1 and 2 refer to carnitine acyltransferases I and II, respectively.*

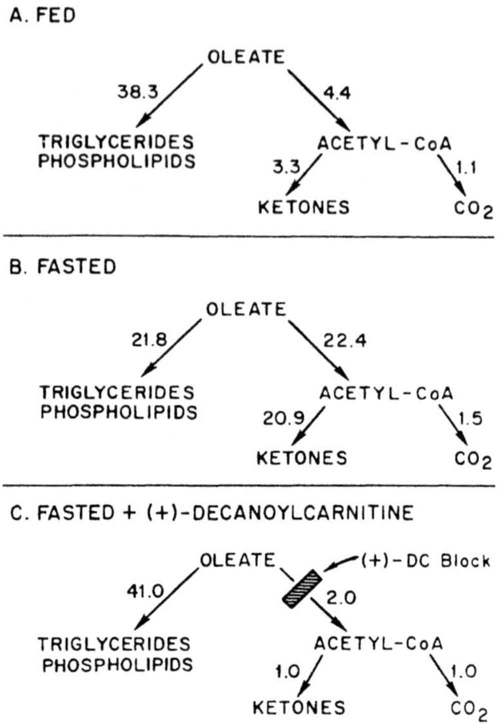


*Figure 6. Metabolism of oleic acid in perfused livers from fed and fasted rats. Values refer to the flux of oleic acid through various pathways and for simplicity are expressed in arbitrary units. (+)-Decanoylcarnitine was used to inhibit the carnitine acyltransferase reaction. For full details see McGarry and Foster (1971b); McGarry et al. (1973).*

the operation of one of these mechanisms to the exclusion of the other we have tended to favor the latter as a primary event on the basis of the following observations. First, studies in rats have shown a poor correlation between ketogenic capacity and the liver content of α-glycerophosphate (McGarry and Foster, 1971b; Williamson, Veloso, Ellington, and Krebs, 1969), a compound previously considered to be a controlling factor for fatty acid esterification (Mayes and Felts, 1967; Tsur *et al.*, 1964). Second, livers from severely ketotic diabetic animals are invariably found to be engorged with fat (Chernick and Scow, 1959; Heimberg, Van Harken, and Brown, 1967; Meier *et al.*, 1972), a feature difficult to reconcile with a defective esterification system in ketotic states. Third, when oxidation of long-chain fatty acids in the ketotic liver was blocked by (+)-acylcarnitines (inhibitors of carnitine acyltransferase) the pattern of fatty acid metabolism immediately reverted to that seen in normal liver, *i.e.*, the substrate was almost completely esterified (Figure 6C). Taken together, these findings support the concept that primary control over the ultimate fate of fatty acids in liver is exerted at some point in the oxidation sequence. We considered that a logical candidate for such control would be the first step specific to long-chain fatty acid oxidation, the carnitine acyltransferase reaction (Figure 5). Consistent with this thesis was the finding that octanoic acid, which enters the mitochondrion by a carnitine-independent mechanism, was oxidized to acetyl-CoA at similar rates in livers from normal and ketotic animals (Figure 7). Although ketogenesis from octanoate was still more rapid in the ketotic compared with the normal liver, this resulted from altered disposal of acetyl-CoA and <u>not</u> from a difference in the generation of the $C_2$ unit, as in the case of long-chain fatty acids (Figures 5 and 6). In contrast, the medium chain fatty acid could be made to act like a long-chain fatty acid when presented to the liver in the form of its carnitine ester. As seen from Table 1, ketone production from octanoylcarnitine, which requires the operation of carnitine acyltransferase II prior to its oxidation, paralleled almost exactly that from long-chain fatty acids.

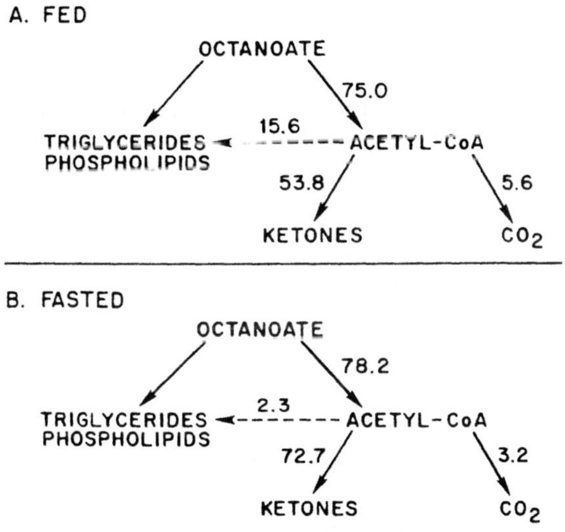


*Figure 7. Metabolism of octanoic acid in perfused livers from fed and fasted rats. Values refer to the flux of octanoic acid through various pathways and for simplicity are expressed in arbitrary units. For full details see McGarry and Foster (1971a).*

*Table 1. EFFECTS OF ANTI-INSULIN SERUM, GLUCAGON, AND FASTING ON HEPATIC KETOGENESIS FROM (-)-OCTANOYLCARNITINE*

| Treatment of animals | Ketone body production from: Octanoylcarnitine (A) | Endogenous fatty acids (B) | A/B |
|---|---|---|---|
| None | 5.3 ± 0.7 | 21 ± 2.8 | 0.25 |
| AIS, 3 hr | 24.6 ± 1.3 | 102 ± 3.0 | 0.24 |
| Glucagon, 3 hr | 23.5 ± 0.8 | 96 ± 5.5 | 0.24 |
| Fasted, 24 hr | 34.4 ± 1.3 | 114 ± 2.3 | 0.30 |

*Fed rats were treated for 3 hours with anti-insulin serum (AIS) or glucagon as described in Figure 2 or were fasted for 24 hours. The livers were then perfused with (-)-[1-$^{14}$C] octanoylcarnitine (1.25 mM). Values are means ± SEM for 4-5 livers in each group and represent the production of ketone bodies (µmol. 100g body wt.$^{-1}$.45 min.$^{-1}$) from the labeled substrate and from endogenous fatty acids. For full details see McGarry and Foster (1974); McGarry et al. (1975).*

## B. Relationship between Liver Glycogen, Carnitine, and Ketogenic Capacity

The precise mechanism whereby circumstances of glucagon excess stimulate the flow of fatty acids through the carnitine acyltransferase step remains a perplexing problem. Since the absolute amount of the enzyme is unchanged in livers from normal and ketotic animals (DiMarco and Hoppel, 1975) and no natural metabolite has, so far, been found to inhibit its activity (Mannaerts, McGarry, and Foster, 1976), it appears that the regulatory system *in vivo* is of a nature that is lost upon disruption of the liver. In seeking to clarify this issue we were struck by the inverse relationship between the liver's content of glycogen and its ketogenic capacity, irrespective of the treatment employed to alter the latter parameter (Figure 8). The correlation was all the more intriguing in view of the fact that the most potent physiological action of glucagon known is its ability to stimulate hepatic glycogenolysis. It, thus, seemed possible that glycogen itself, or more likely a related liver component whose concentration fluctuates in parallel with that of glycogen, might influence the efficiency of the carnitine acyltransferase step. However, when added to the medium perfusing livers from fed rats, glucagon failed to stimulate ketogenesis despite its obvious glycogenolytic effect (Figure 9). While such *in vitro* experiments do not necessarily preclude a role for glycogen (or related compound) in the control of hepatic fatty acid metabolism they suggested that other factors must also be involved. One such factor appears to be the tissue concentration of carnitine,

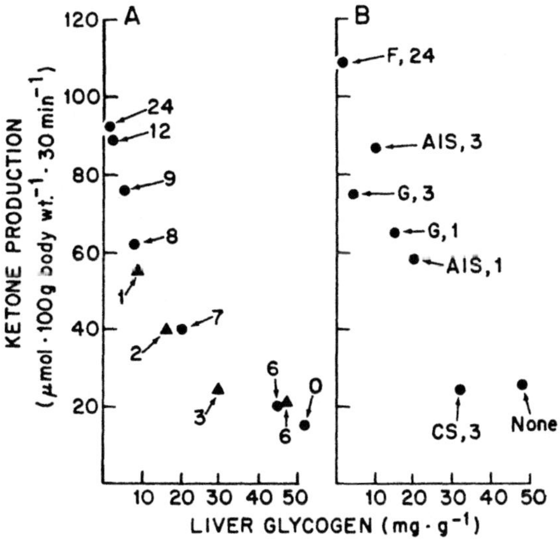


*Figure 8. Relationship between glycogen content and ketogenic capacity of rat liver. Panel A: (●) fed rats fasted for the indicated number of hours; (▲) 24 hour fasted rats refed a high-glucose diet for the indicated number of hours. Panel B: fed rats infused with control guinea pig serum containing no additions (CS), insulin antibodies (AIS), or glucagon (G) or fasted (F) for the indicated number of hours. Values are for 4-8 animals in each group. Standard errors have been omitted for purposes of clarity. For full details see McGarry et al. (1973); McGarry et al. (1975).*

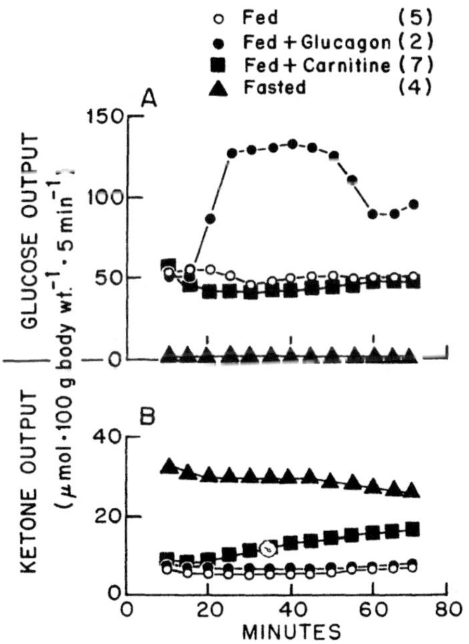


*Figure 9. Effects of glucagon and carnitine on ketogenesis from oleate in perfused livers from fed rats. Livers were perfused in non-recirculating fashion with medium containing 0.7 mM oleic acid. At the 15-minute time point an infusion of glucagon or L-carnitine was begun to give final concentrations of 1 ng per ml and 0.5 mM, respectively. Values are means for the numbers of animals shown in parentheses. Standard errors have been omitted for reasons of clarity. For further details see McGarry, Robles-Valdes, and Foster (1975).*

a substrate for the carnitine acyltransferase I reaction. This follows from the observation that when added to the perfusion medium carnitine stimulated ketogenesis from oleate in livers from fed rats (Figure 9). In addition, a variety of manipulations used to turn on the ketogenic system *in vivo* (short-term treatment with insulin antibodies or glucagon, starvation, and induction of experimental diabetes) also resulted in elevation of the liver carnitine content (Table 2). Thus, all four treatment schedules, each of which produced an increase in the blood [glucagon]:[insulin] ratio, caused a marked depletion of glycogen, elevation of carnitine and concomitant stimulation of fatty acid oxidation in the liver.

To gain further insight into these relationships we carried out a series of experiments with the neonatal rat. Our reasons were as follows. In this species the transition from intrauterine to extrauterine life is accompanied by an abrupt fall in plasma insulin and a rise in plasma glucagon concentration (Girard, Cuendet, Marliss, Kervran, Rieutort, and Assan, 1973). This alteration in hormonal balance coincides with a number of metabolic adaptations that simulate those seen during the induction of ketosis in the adult animal:fat (from maternal milk) replaces glucose (from maternal blood) as the major respiratory fuel; liver glycogen stores are rapidly dissipated and marked ketonemia supervenes (Bailey and Lockwood, 1973; Page, Krebs, and Williamson, 1971). This represents a further example of physiological ketosis in which acetoacetate and β-hydroxybutyrate appear to be crucial for normal development of brain and probably other tissues. From the standpoint of the present discussion it was of interest to find that in this situation also the concentration of liver carnitine paralleled the

*Table 2. RELATIONSHIP BETWEEN CARNITINE CONTENT AND KETOGENIC CAPACITY*

| State of animal | Free carnitine[a] | Total carnitine[a] | Ketone production from oleate[b] |
|---|---|---|---|
| Fed | 64 ± 7 | 103 ± 7 | 26 ± 3 |
| Fasted, 24 hr | 220 ± 17 | 287 ± 27 | 118 ± 8 |
| Diabetic | 355 ± 33 | 487 ± 49 | 192 ± 13 |
| Fed, glucagon | 130 ± 13 | 269 ± 21 | 87 ± 5 |
| Fed, AIS | 81 ± 6 | 224 ± 9 | 87 ± 2 |

*Values are means ± SEM for 6 animals in each group.*

[a] *nmol.g wet wt.*$^{-1}$
[b] *μmol.g body wt.*$^{-1}$ *30 min.*$^{-1}$

*Data taken from McGarry, Robles-Valdes, and Foster (1975).*

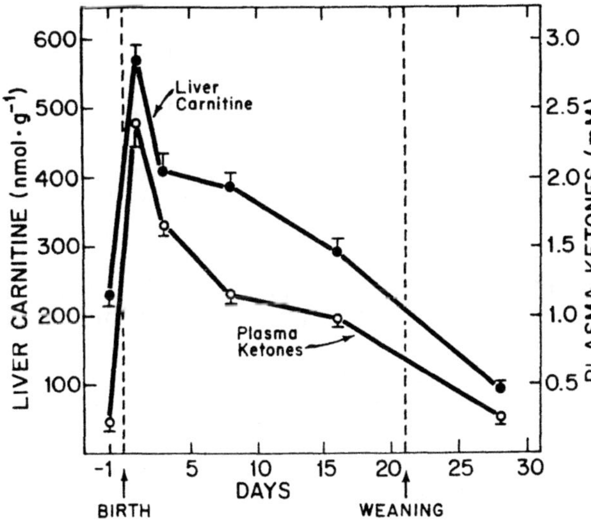


*Figure 10. Liver carnitine (free + esterified) and plasma ketone concentrations in the developing rat. Values are means ± SEM for pooled samples from 4-8 litters at each time point. For full details see Robles-Valdes et al. (1976).*

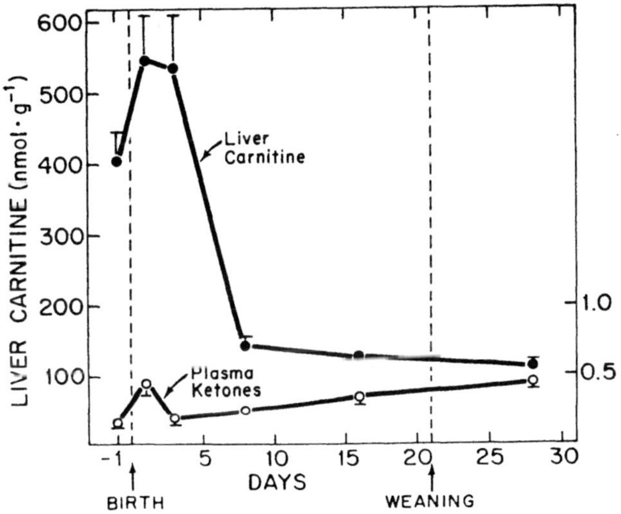


*Figure 11. Liver carnitine (free + esterified) and plasma ketone concentrations in maternal rats during late gestation and nursing. Values are means ± SEM for 4-6 animals at each time point. For full details see Robles-Valdes et al. (1976).*

rate of ketone production, assuming that the latter is reflected by the blood ketone level (Figure 10). However, when similar studies were carried out with nursing mother rats the unexpected findings shown in Figure 11 were obtained. Liver carnitine levels again increased in late gestation, reached peak values at days 1-3 postpartum and returned to normal during the next five days; yet throughout this period the blood ketone concentration remained low. Since these animals were consuming a high carbohydrate diet, plasma insulin levels were presumably adequate to restrain adipose tissue lipolysis, accounting for the low plasma FFA levels (data not shown) and the absence of ketonemia. But why should the maternal liver be enriched with carnitine under conditions where a requirement for active ketogenesis was absent? Further studies, described in detail elsewhere (Robles-Valdes *et al.*, 1976), provided a plausible answer when it was shown that the maternal liver actively synthesizes carnitine for transport via milk to the suckling pups. Presumably the neonatal liver is dependent upon this mechanism to assure rapid rates of hepatic fatty acid oxidation and ketogenesis.

The question still remained as to whether, in accord with our previous findings, the increased carnitine concentration of the maternal liver would impart upon the organ a high ketogenic capacity when perfused *in vitro* with adequate fatty acid substrate. When this was tested experimentally the results were again surprising. As seen from Table 3, in the fed state ketone production by the maternal livers was modestly elevated over that seen in livers from

*Table 3. KETOGENIC CAPACITY, CARNITINE, AND GLYCOGEN CONTENT OF FEMALE RAT LIVER*

| Type of rat | Ketone body production[a] | Liver carnitine[b] | Liver glycogen[c] |
|---|---|---|---|
| Virgin females | | | |
| Fed | 2.0 ± 0.6 | 160 ± 16 | 55.0 ± 1.6 |
| Fasted | 28.8 ± 1.5 | 361 ± 34 | 0.9 ± 0.4 |
| Nursing mothers | | | |
| Fed | 6.0 ± 2.6 | 417 ± 60 | 52.6 ± 4.1 |
| Fasted | 26.8 ± 1.9 | 487 ± 26 | 0.8 ± 0.3 |

*Values are means ± SEM for 4 animals in each group.*

[a] *μmol.g liver$^{-1}$ 30 min.$^{-1}$ on perfusion with oleic acid*

[b] *nmol.g wet wt.$^{-1}$*

[c] *mg.g wet wt.$^{-1}$*

*For full details see Robles-Valdes et al. (1976).*

virgin control rats, a feature probably related to the difference in carnitine concentration in the two groups. However, both rates were low when compared with the values found after a 24-hour fast. While fasting produced the expected increase in liver carnitine in the control animals, it caused essentially no change in the already high level in the maternal liver. It must, therefore, be concluded that carnitine enrichment *per se* is not sufficient to assure maximal hepatic ketogenic capacity. While the solution to the paradox is not yet clear, it is noteworthy that, in contrast to all of our previous studies which utilized either male adult rats or neonatal animals and in which enhanced ketogenesis was accompanied both by increased liver carnitine and decreased liver glycogen levels (Figures 8 and 10), these two components were simultaneously elevated in the livers from nursing mother rats. It seems likely, therefore, that both an increase in carnitine concentration and depletion of glycogen are required for full induction of hepatic ketogenic potential. A clinical counterpart to these observations in the rat may be found in the studies of Havel, Balasse, Williams, Kane, and Segel (1969) and of Binkiewicz and Senior (1973) in subjects with Von Gierke's disease (Glycogenosis Type 1). On fasting, such patients displayed the expected increase in plasma free fatty acid concentrations but much lower blood ketone levels than did control individuals, indicating a marked reduction in hepatic ketogenic capacity.

As noted earlier, fluctuations in liver glycogen with attendant perturbations of carbohydrate metabolism in response to changing blood concentrations of insulin and glucagon are understandable on the basis of current concepts of the actions of these two hormones. What remains to be clarified is the precise link between this aspect of liver metabolism and the first step in the fatty acid oxidation sequence. Other questions for the future concern the mechanism by which elevation of the [glucagon]:[insulin] ratio brings about an increase in liver carnitine content, and how the same effect is achieved in the nursing mother rat under circumstances where plasma insulin and glucagon levels are presumably within the normal range.

## V. CONCLUDING REMARKS

It is evident from the foregoing discussion that the century-old question of how the ketogenic process is controlled is by no means solved. Nevertheless, important advances have been made in recent years, central among which has been the increasing recognition that the development of the ketotic state involves metabolic adaptations in two organ systems, adipose tissue and liver. Alterations at both sites are essential and both must be considered in dealing with the overall question of control mechanisms. While this might entail interactions between many hormones, we have emphasized two that appear to play prominent roles, namely insulin and glucagon. In this regard, it is noteworthy that glucagon excess has been clearly implicated by Unger and associates (see Unger, 1976) as a key factor in the etiology of diabetic hyperglycemia. It is now tempting to speculate that the other major complication of uncontrolled diabetes, ketoacidosis, is likewise the result of glucagon's actions in a setting of insulin deficiency.

## ACKNOWLEDGMENTS

The authors' studies described here were supported by United States Public Health Service Grant AM18575, Training Grant CA05200 and a grant from the American Diabetes Association. Dr. McGarry is the recipient of Research Career Development Award 1-K04-AM0763.

## REFERENCES

Alberti, K.G.M.M., Hockaday, T.D.R., and Turner, R.C. (1973). Small doses of intramuscular insulin in the treatment of diabetic "coma." *Lancet* II:515-522.

Armin, J.R., Grant, T., and Wright, P.H. (1960). Acute insulin deficiency provoked by single injections of anti-insulin serum. *J. Physiol.* (Lond.) 153:131-145.

Bailey, E. and Lockwood, E.A. (1973). Some aspects of fatty acid oxidation and ketone body formation and utilization during the development of the rat. *Enzyme* 15:239-253.

Balasse, E.O., Bier, D.M., and Havel, R.J. (1972). Early effects of anti-insulin serum on hepatic metabolism of free fatty acids in dogs. *Diabetes* 21:280-288.

Binkiewicz, A. and Senior, B. (1973). Decreased ketogenesis in Von Gierke's disease (Type 1 glycogenosis). *J. Peds.* 83:973-978.

Björntorp, P. and Scherstén, T. (1967). Effect of β-hydroxybutyrate on lipid mobilization. *Amer. J. Physiol.* 212:683-687.

Chernick, S.S. and Scow, R.O. (1959). Early effects of 'total' pancreatectomy on fat metabolism in the rat. *Amer. J. Physiol.* 196:125-131.

DiMarco, J.P. and Hoppel, C. (1975). Hepatic mitochondrial function in ketogenic states. Diabetes, starvation and after growth hormone administration. *J. Clin. Invest.* 55:1237-1244.

Engel, F.L. and Amatruda, T.T., Jr. (1963). Hormonal aspects of ketosis. *Ann. N.Y. Acad. Sci.* 104:753-771.

Gerich, J.E., Lorenzi, M., Bier, D.M., Schneider, V., Tsalikian, E., Karam, J.H., and Forsham, P.H. (1975). Prevention of human diabetic keotacidosis by somatostatin. Evidence for an essential role of glucagon. *N. Eng. J. Med.* 292, 985-989.

Girard, J.R., Cuendet, G.S., Marliss, E.B., Kervran, A., Rieutort, M., and Assan, R. (1973). Fuels, hormones, and liver metabolism at term and during the early postnatal period in the rat. *J. Clin. Invest.* 52:3190-3198.

Havel, R.J., Balasse, E.O., Williams, E.E., Kane, J.P., and Segel, N. (1969). Splanchnic metabolism in Von Gierke's disease (Glycogenosis Type 1). *Trans. Assoc. Amer. Phys.* 82:305-323.

Heimberg, M., Van Harken, D.R., and Brown, T.O. (1967). Hepatic lipid metabolism in experimental diabetes. I. Incorporation of [1-$^{14}$C] palmitate into lipids of the liver and of the $d<$ 1.020 perfusate lipoproteins. *Biochim. Biophys. Acta* 137: 435-445.

Houssay, B.A. (1963). Hormonal factors of diabetic ketosis. *Diabetes* 12:481-489.

Krebs, H.A. (1966). The regulation of the release of ketone bodies by the liver. *Adv. Enzyme Regul.* 4:339-353.

Krebs, H.A., Williamson, D.H., Bates, M.W., Page, M.A., and Hawkins, R.A. (1971). The role of ketone bodies in caloric homeostasis. *Adv. Enzyme Regul.* 9:387-409.

Levy, L.J., Duga, J., Gingis, M., and Gordon, E. (1973). Ketoacidosis associated with alcoholism in non-diabetic subjects. *Ann. Int. Med.* 78:213-219.

Lossow, W.J., Brown, G.W., Jr., and Chaikoff, I.L. (1956). The action of insulin in sparing fatty acid oxidation: A study with palmitic acid-1-C$^{14}$. *J. Biol. Chem.* 220:839-849.

Lossow, W.J. and Chaikoff, I.L. (1954). Carbohydrate sparing of fatty acid oxidation. I. The relation of fatty acid chain length to the degree of sparing. II. The mechanism by which carbohydrate spares the oxidation of palmitic acid. *Arch. Biochem. Biophys.* 57:23-40.

Madison, L.L., Mebane, D., Unger, R.H., and Lochner, A. (1964). The hypoglycemic action of ketones. II. Evidence for a stimulatory feedback of ketones on the pancreatic beta cells. *J. Clin. Invest.* 43:408-415.

Mannaerts, G., McGarry, J.D., and Foster, D.W. (1976). Unpublished observations.

Mayes, P.A. and Felts, J.M. (1967). The regulation of fat metabolism in the liver. *Nature* (Lond.) 215:716-718.

McGarry, J.D. and Foster, D.W. (1971a). The regulation of ketogenesis from octanoic acid. The role of the tricarboxylic acid cycle and fatty acid synthesis. *J. Biol. Chem.* 246:1149-1159.

McGarry, J.D. and Foster, D.W. (1971b). The regulation of ketogenesis from oleic acid and the influence of antiketogenic agents. *J. Biol. Chem.* 246:6247-6253.

McGarry, J.D. and Foster, D.W. (1974). The metabolism of (-)-octanoylcarnitine in perfused livers from fed and fasted rats. Evidence for a possible regulatory role of carnitine acyltransferase in the control of ketogenesis. *J. Biol. Chem.* 249: 7984-7990.

McGarry, J.D., Meier, J.M., and Foster, D.W. (1973). Effects of starvation and refeeding on carbohydrate and lipid metabolism *in vivo* and in the perfused rat liver. *J. Biol. Chem.* 248: 270-278.

McGarry, J.D., Robles-Valdes, C., and Foster, D.W. (1975). Role of carnitine in hepatic ketogenesis. *Proc. Nat'l. Acad. Sci.* 72: 4385-4388.

McGarry, J.D., Wright, P.H., and Foster, D.W. (1975). Hormonal control of ketogenesis. Rapid activation of hepatic ketogenic capacity in fed rats by anti-insulin serum and glucagon. *J. Clin. Invest.* 55:1202-1209.

Meier, J.M., McGarry, J.D., and Foster, D.W. (1972). Studies of the development of diabetic ketosis in the rat. *J. Lip. Res.* 13:228-233.

Ontko, J.A. (1972). Metabolism of free fatty acids in isolated liver cells. *J. Biol. Chem.* 247:1788-1800.

Owen, O.E., Felig, P., Morgan, A.P., Wahren, J., and Cahill, G.F., Jr. (1969). Liver and kidney metabolism during prolonged starvation. *J. Clin. Invest.* 48:574-583.

Owen, O.E., Morgan, A.P., Kemp, H.G., Sullivan, J.M., Herrera, M.G., and Cahill, G.F., Jr. (1967). Brain metabolism during fasting. *J. Clin. Invest.* 46:1589-1595.

Page, M.A., Krebs, H.A., and Williamson, D.H. (1971). Activities of enzymes of ketone body utilization in brain and other tissues of suckling rats. *Biochem. J.* 121:49-53.

Robles-Valdes, C., McGarry, J.D., and Foster, D.W. (1976). Maternal-fetal carnitine relationships and neonatal ketosis in the rat. *J. Biol. Chem.* 251:6007-6012.

Schade, D.S. and Eaton, R.P. (1975). Glucagon regulation of plasma ketone body concentration in human diabetes. *J. Clin. Invest.* 56:1340-1344.

Scow, R.O. and Chernick, S.S. (1960). Hormonal control of protein and fat metabolism in the pancreatectomized rat. *Recent Progr. Hormone Res.* 16:497-545.

Tzur, R., Tal, E., and Shapiro, B. (1964). α-Glycerophosphate as a regulatory factor in fatty acid esterification. *Biochim. Biophys. Acta* 84:18-23.

Unger, R.H. (1976). Diabetes and the alpha cell. *Diabetes* 25: 136-151.

Williamson, J.R. (1967). Effects of fatty acids, glucagon and anti-insulin serum on the control of gluconeogenesis and ketogenesis in rat liver. *Adv. Enzyme Regul.* 5:229-255.

Woodside, W.F. and Heimberg, M. (1972). Hepatic metabolism of free fatty acids in experimental diabetes. *Israel. J. Med. Sci.* 8:309-316.

Woodside, W.F. and Heimberg, M. (1976). Effects of anti-insulin serum, insulin and glucose on output of triglycerides and on ketogenesis by the perfused rat liver. *J. Biol. Chem.* 251: 13-23

# INFLUENCE OF INSULIN AND GLUCAGON ON KETOGENESIS BY ISOLATED RAT HEPATOCYTES[1]

William F. Woodside

Department of Pharmacology
University of New Mexico School of Medicine
Albuquerque, NM 87131

## INTRODUCTION

The quantitative roles of insulin and glucagon in the regulation of hepatic ketogenesis are still not defined adequately. Glucagon added *in vitro* has been reported by certain workers to stimulate ketogenesis in perfused livers from normal-fed rats (Heimberg, Weinstein, and Kohout, 1969; Weinstein, Klausner, and Heimberg, 1973). However, other workers did not observe a ketogenic effect of the hormone under similar conditions (McGarry and Foster, 1978; McGarry, Wright, and Foster, 1975).

Similar studies with insulin have also given conflicting results. Insulin added *in vitro* was reported to suppress ketogenesis in livers from normal-fed rats perfused with oleate but the quantitative changes were small (Heimberg, Van Harken, Weinstein, and Kohout, 1969). The hormone added *in vitro* also suppressed ketogenesis in perfused livers from rats pretreated with anti-insulin serum (AIS) for 3 *h* (Woodside and Heimberg, 1976). Insulin alone had no effect, however, on ketogenesis in perfused livers from rats pretreated with AIS for 10 *h*, although the hormone did suppress this process when glucose was present in the perfusate (Woodside and Heimberg, 1976). Other workers have reported that the hormone had no antiketogenic effect in perfused livers from fasted (McGarry and Foster, 1971), alloxan diabetic (Heimberg *et*

[1]This work was supported, in part, by grants from New Mexico Heart Association, Pharmaceutical Manufacturers Association, and the New Mexico Medical Foundation.

*al.*, 1969), or pancreatectomized (Scow and Chernick, 1960) rats even when glucose was added to the perfusate.

Thus, it is clear that further study is necessary to define the quantitative roles of insulin and glucagon in the direct regulation of hepatic ketogenesis. We have studied this problem utilizing isolated rat hepatocytes from normal fed rats and are reporting our initial findings here.

## EXPERIMENTAL METHODS AND RESULTS

Isolated hepatocytes were prepared from the livers of normal fed male rats weighing 200-300 *gms* by perfusion with collagenase to digest the connective tissue. The basic procedures used were similar to those described by Seglen (1973) with the exception that Krebs-Ringer bicarbonate (KRB) solution was used as the basic perfusion medium instead of the balanced salt solution buffered with HEPES. Isolated cells were suspended at a concentration of 25-35 *mg/mℓ* in KRB containing 4% bovine serum albumin (BSA) which had been defatted and dialyzed prior to use. The suspended cells were preincubated while being gassed constantly with 95% $O_2$/5% $CO_2$ for 30 *min* prior to use, for reasons described by Claus, Pilkis, and Park (1975). Then 50-70 *mg* cells (wet weight) were added to 25 *mℓ* flasks in a final volume of 4 *mℓ* of 4% BSA in KRB containing variable amounts of oleate and hormone as indicated. The flasks were gassed briefly with 95% $O_2$/5% $CO_2$, stoppered, and incubated for 1 *h*. The reaction mixture was then deproteinized by addition of $Ba(OH)_2$ and $ZnSO_4$. The concentrations of acetoacetate and β-hydroxybutyrate in the supernatant were analyzed by an enzymatic procedure (McGarry, Guest, and Foster, 1970).

The influence of varying doses of glucagon on ketogenesis was examined in isolated hepatocytes from normal-fed rats incubated with either no added fatty acid or with 0.75 *mM* oleate [the approximate concentration of FFA observed in plasma of fasted rats (Scow and Chernick, 1970)]. The results shown in Figure 1 are from a representative experiment, but have been verified in several additional experiments. It can be seen that glucagon in the doses added had no stimulatory effect on ketogenic rates obtained in hepatocytes incubated without added FFA. With 0.75 *mM* oleate, basal rates of ketogenesis were increased approximately two- to three-fold. Moreover, under these conditions, glucagon stimulated ketogenesis markedly. Maximal increases in rate were induced by glucagon at a concentration of 2-3 *nM* and were over four-fold greater than control rates. The concentration of hormone required for one-half maximal stimulation in rate was approximately 0.3 *nM*.

The effects of varying doses of insulin on the stimulation in ketogenic rates induced by 0.25 and 0.50 *nM* glucagon were examined,

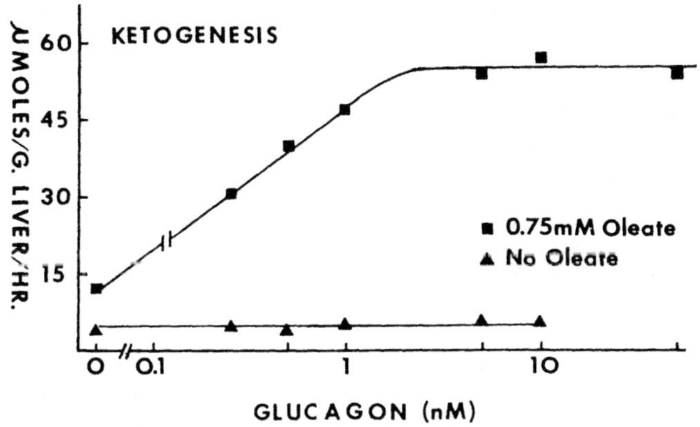


*Figure 1. Influence of glucagon on ketogenesis by isolated hepatocytes from normal-fed rats. Isolated hepatocytes were incubated with varying doses of glucagon in the presence or absence of 0.75 mM oleate, as described in the text. Rates of ketogenesis are expressed as the sum of acetoacetate and β-hydroxybutryrate/g of wet weight of cells.*

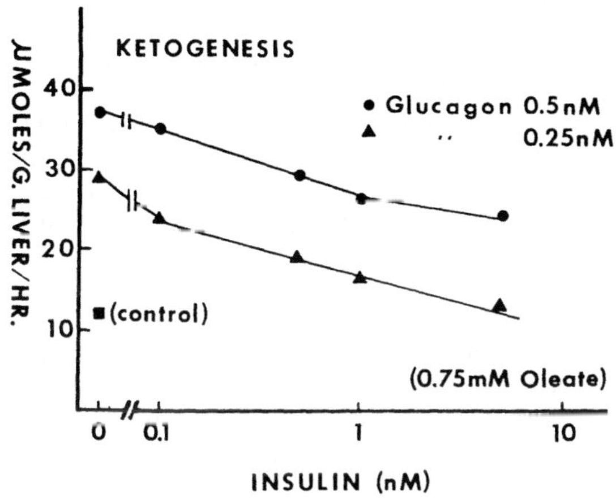


*Figure 2. Interaction of insulin and glucagon in the regulation of ketogenesis by isolated hepatocytes from normal-fed rats. Isolated hepatocytes were incubated with 0.75 mM oleate and 0.25 or 0.50 nM glucagon. Cells were preincubated for 10 min with varying doses of insulin prior to addition of oleate and glucagon. Other details are described in the legend to Figure 1 and in the text.*

and the results are shown in Figure 2. Under these conditions, insulin suppressed ketogenesis in a dose-dependent, nearly linear fashion. Wiht 0.25 *nM* glucagon, insulin at a concentration as low as 0.1 *nM* suppressed ketogenic rates. With an insulin concentration of 5 *nM*, the rate was decreased by 54% to a level which was almost the same as the rate observed without glucagon. With 0.5 *nM* glucagon, insulin suppressed ketogenesis similarly. With 5 *nM* insulin, the rate was reduced by 35%. It is important to note additionally that little suppressive effects of insulin on basal rates of ketogenesis (*i.e.*, rates obtained with 0.75 *mM* oleate alone) were obtained under our experimental conditions.

## DISCUSSION

On the basis of our findings, we would conclude that insulin and glucagon do play a significant quantitative role in the direct regulation of hepatic ketogenesis, in contrast to other workers who concluded the opposite based on their own observations (McGarry and Foster, 1978; McGarry *et al.*, 1975; Exton and Park, 1972). Exton and Park (1972) reported that glucagon added *in vitro* did not exert major stimulatory effects on ketogenesis in perfused livers from normal-fed rats. However, no exogenous fatty acid was added to the perfusate. The data presented in Figure 1 suggest clearly that exogenous FFA must be present in order for optimal quantitative effects of the hormone to be manifest. Indeed, it seems likely that there must be an optimal concentration of FFA at which glucagon can exert its maximal quantitative ketogenic effect, as reported earlier by Heimberg *et al.* (1969) with the perfused liver. McGarry and coworkers (1975, 1978), however, did not observe direct ketogenic effects of glucagon in perfused livers from normal fed rats even when exogenous FFA was added to the perfusate. It seems logical, therefore, that other factors besides the concentration of FFA may influence the ability of glucagon, as well as insulin, to regulate this process directly. Such factors may include the presence of antiketogenic substrates, *e.g.*, alanine, lactate, and glucose in the media, the intrahepatic concentration of glycogen, etc. We are presently carrying out additional studies to examine these various possibilities.

## REFERENCES

Claus, T.H., Pilkis, S.J., and Park, C.R. (1975). Stimulation by glucagon of the incorporation of U-$C^{14}$-labelled substrates into glucose by isolated hepatocytes from fed rats. *Biochim. Biophys. Acta* 404:110-123.

Exton, J.H. and Park, C.R. (1972). Interaction of insulin and glucagon in the control of liver metabolism. pp. 437-455. In R.O. Greep and E.B. Astwood (Eds.) *Handbook of Physiology*, Section 7, *Endocrinology*. American Physiological Society, Washington, DC.

Heimberg, M., Van Harken, D.R., Weinstein, I., and Kohout, M. (1969). Regulatory factors in ketogenesis and in the metabolism of triglycerides by liver. *Adv. Exp. Biol. Med.* 4:185-199.

Heimberg, M., Weinstein, I., and Kohout, M. (1969). The effects of glucagon, dibutyryl cyclic adenosine 3',5'-monophosphate and concentration of free fatty acid on hepatic lipid metabolism. *J. Biol. Chem.* 244:5131-5139.

McGarry, J.D. and Foster, D.W. (1971). The regulation of ketogenesis by oleic acid and the influence of anti-ketogenic agents. *J. Biol. Chem.* 246:6427-6253.

McGarry, J.D. and Foster, D.W. (1978). Hormonal control of ketogenesis. pp. 79-96. In D.M. Klachko, R.R. Anderson, and M. Heimberg (Eds.) *Hormones and Energy Metabolism*. Plenum Press, New York.

McGarry, J.D., Guest, M.J., and Foster, D.W. (1970). Ketone body metabolism in the ketosis of starvation and alloxan diabetes. *J. Biol. Chem.* 245:4382-4390.

McGarry, J.D., Wright, P.H., and Foster, D.W. (1975). Hormonal control of ketogenesis. Rapid activation of hepatic ketogenic capacity in fed rats by anti-insulin serum and glucagon. *J. Clin. Invest.* 55:1202-1209.

Seglen, P.O. (1973). Preparation of liver cells. III. Enzymatic requirements for tissue dispersion. *Exp. Cell Res.* 82:391-398.

Scow, R.O. and Chernick, S.S. (1960). Hormonal control of protein and fat metabolism in the pancreatectomized rat. *Rec. Progr. Horm. Res.* 16:497-541.

Scow, R.O. and Chernick, S.S. (1970). Mobilization, transport, and utilization of free fatty acids. *Comp. Biochem.* 18:19-49.

Weinstein, I., Klausner, H.A., and Heimberg, M. (1973). The effect of concentration of glucagon on output of triglyceride, ketone bodies, glucose and urea by the liver. *Biochim. Biophys. Acta* 300:300-309.

Woodside, W.F. and Heimberg, M. (1976). Effects of anti-insulin serum, insulin, and glucose on output of triglycerides and ketogenesis by the perfused rat liver. *J. Biol. Chem.* 251:13-23.

## ACKNOWLEDGEMENTS

The expert technical assistance of Sherry Whitehurst in the performance of this work is gratefully acknowledged.

# HORMONAL CONTROL OF GLYCOGEN METABOLISM[1]

J. Larner, P.J. Roach, L.C. Huang, G. Brooker,
F. Murad, and R. Hazen

Department of Pharmacology
University of Virginia
Charlottesville, VA 22903

## OUTLINE

I. Introduction

II. Basic concepts of the regulation of glycogen synthesis and degradation

III. Altered kinase and phosphatase activity with insulin action

IV. The insulin intermediate--"X"

V. Summary and conclusions

## INTRODUCTION

It is now recognized that both glycogen synthesis and degradation are controlled by covalent modification as well as by soluble cellular effectors. In general, it appears that only the hormonal controls are exerted through covalent phosphorylation and dephosphorylation. By this mechanism one cell type, the endocrine cell, communicates with another cell type, the target cell. The soluble cellular effectors appear to be influenced by nonhormonal as well as hormonal stimuli.

## BASIC CONCEPTS OF THE REGULATION OF GLYCOGEN SYNTHESIS AND DEGRADATION

Figure 1 diagrams the hormonal events in outline form. The

[1]This work was supported by U.S. Public Health Service Grants No. 2 R01 AM15334-08 and AM17043-01.

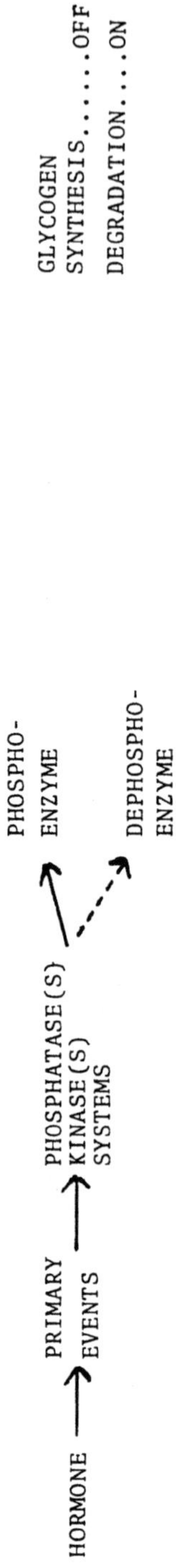


*Figure 1. Hormone cascade-general.*

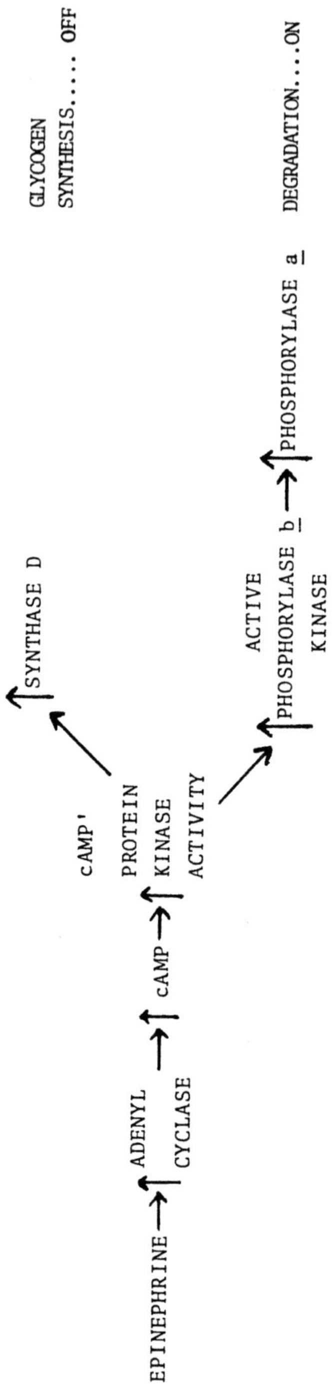


*Figure 2. Hormone cascade-epinephrine.*

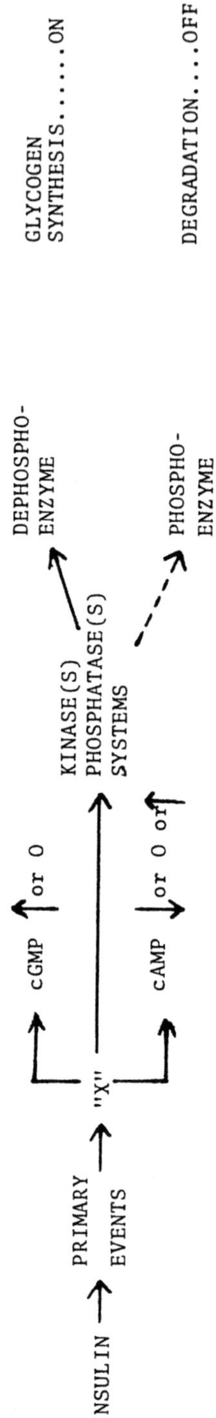


*Figure 3. Hormone cascade-insulin.*

hormone initiates a set of primary events which, in turn, leads to a modification of the interconverting kinase-phosphatase systems. This, in a cascading series of reactions, alters the state of phosphorylation of the enzymes glycogen synthase and phosphorylase. As written, the phosphorylation system activates the mobilization of glycogen. If we take epinephrine, as an example, the events can be written more specifically as shown in Figure 2. Adenylate cyclase is stimulated to form increased cyclic AMP, activating the cyclic AMP-dependent protein kinase which leads to the phosphorylation of glycogen synthase and of phosphorylase _b_ kinase. The latter enzyme phosphorylates phosphorylase to its active or _a_ form and, as a result, glycogen synthesis is decreased and degradation is increased.

In the case of insulin, we propose the series of events depicted in Figure 3. According to this hypothesis, insulin acts first to produce its unique mediator, called "X." "X" affects the kinase (Larner, Huang, Brooker, Murad, and Miller, 1974) and/or phosphatase systems, which results in the enhanced formation of the dephosphorylated, active or _I_ form of glycogen synthase (Larner, Huang, Hazen, Brooker, and Murad, 1975; Larner, Takeda, Brewer, Huang, Hazen, Brooker, Murad, and Roach, 1976). The level of active phosphorylase may either remain unaltered or is decreased depending on the particular tissue (Bishop and Larner, 1967; Craig and Larner, 1964). The tissue levels of cyclic AMP and cyclic GMP are affected by insulin in a variable manner--that is, either increased, decreased, or not affected in the case of cyclic AMP, and either increased or not affected in the case of cyclic GMP. Therefore, we consider that these fluctuations are not in the direct pathway of insulin action but on side pathways also influenced by "X." This scheme, in essence, has been held by us over the years and was presented in 1972 (Larner, 1972). The points to be discussed include (a) the altered kinase and phosphatase activity following insulin and epinephrine or glucagon action, (b) the evidence for the existence of a possible intermediate substance "X," and finally (c) the question of the role of multiple phosphorylation in the control of glycogen synthase as contrasted with single phosphorylation in the control of phosphorylase.

## ALTERED KINASE AND PHOSPHATASE ACTIVITY WITH INSULIN ACTION

Because of the lack of an effect of insulin on basal cyclic AMP levels in muscle and in liver in our laboratory (Larner, 1972), we asked the following question, "Does insulin alter the sensitivity of the kinase to cyclic AMP rather than directly affecting the cellular level of the nucleotide?" In effect, this question consists of "turning the coin on its opposite side," and would produce the same end result as an altered level of cyclic AMP. In 1967, the first experimental evidence for a stable kinase alteration brought on by insulin was reported (Villar-Palasi and Wenger, 1967).

They demonstrated that after injection of insulin into rats *in vivo*, the kinase measured without cyclic AMP (I kinase) was decreased, with no difference in the "total" kinase measured in the presence of cyclic AMP (Figure 4). This suggested that the kinase existed in two forms, and that insulin promoted an interconversion of the independent or I form into the dependent or D form of the enzyme. In 1970, Shen, Villar-Palasi, and Larner showed the same shift of the kinase in diaphragms treated *in vitro* with in-

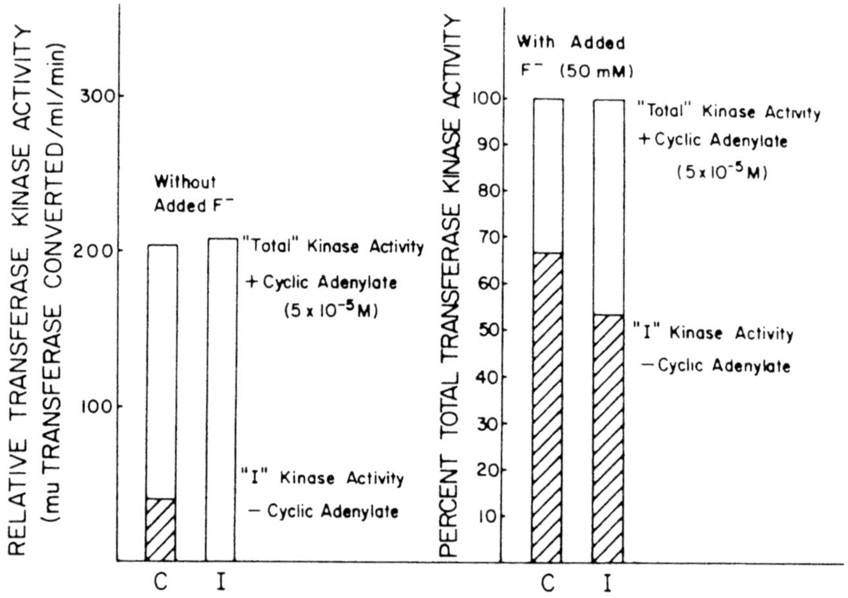


*Figure 4. Transferase kinase activity of extracts of skeletal muscle of rats injected with saline or insulin. Control-C, Insulin=I. Fed male rats were anaesthetized and injected with saline or insulin (2 U/k i.p.). After 20 min, muscle was removed from hind legs, homogenized 1:7 (w/v) in either 50 mM Tris-Cl buffer pH 8.2 containing 5 mM versene or in 50 mM Tris-Cl buffer pH 8.2 containing 5 mM versene and 50 mM KF, and centrifuged. These extracts were assayed for transferase kinase as follows: to each 0.2 ml aliquot of extract, 0.1 ml purified I and either 0.1 ml 30 mM ATP-60 mM $MgCl_2$, 10 mM Tris-Cl pH 8.2 or 0.1 ml 30 mM ATP-60 mM $MgCl_2$ 0.3 mM cyclic AMP, 15 mM caffeine, 20 mM Tris-Cl pH 8.2 was added. Tubes were incubated at 30°C. Transferase was assayed in aliquots at zero time and after 5 or 7 min of kinase action in the presence of 50 mM KF and 20 mM versene to inhibit further phosphatase or kinase action during the transferase assay.*

sulin and, in addition, showed that epinephrine promoted the opposite D to I form shift. In 1971, the I to D shift was shown to occur with perfusion alone in the *in vitro* perfused rat heart (Nuttall and Larner, 1971). In 1972, we again demonstrated the I to D shift in diaphragms treated with insulin *in vitro* and showed that it was blocked by the pituitary diabetogenic peptide of Louis and Conn (Miller and Larner, 1972). In 1973, we next demonstrated the same I to D form shift of the kinase in rat liver perfused *in situ* with insulin, and, in addition, demonstrated the opposite shift of the D to I form of the kinase with glucagon treatment (Miller and Larner, 1973).

From other laboratories, evidence of altered kinase activity with hormone action also began to accumulate. In 1971, it was reported that in rat diaphragms treated with insulin *in vitro*, coincident with the increase in the D form of the kinase, there was an increased binding of cyclic AMP (Walaas, Walaas, and Grønnerød, 1972). This suggested that the kinase was present to a greater extent in the form of the holoenzyme after insulin treatment. The inverse was true after epinephrine treatment. The I form of the enzyme was increased along with a decreased binding of cyclic AMP. The overall mechanism of these hormonal influences can, therefore, be visualized as shown in Figure 5, which demonstrates how these hormonal perturbations of the kinase in intact cells fit the two subunit dissociation-association model of the kinase worked out

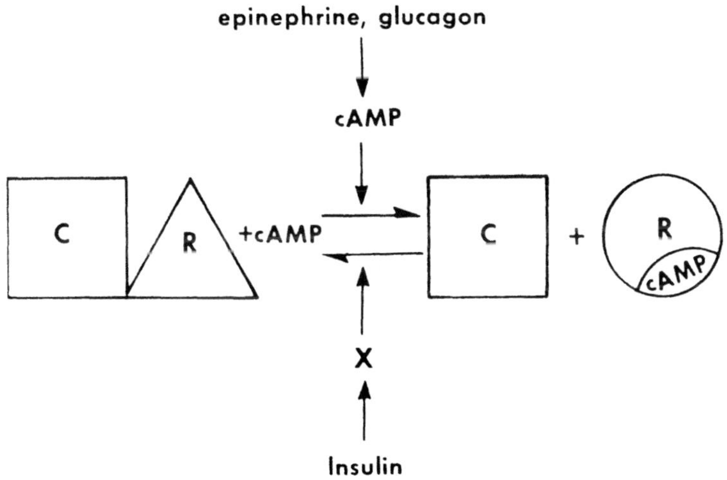


*Figure 5. Dissociation-association model of the cyclic AMP-dependent protein kinase.*

*C = Catalytic subunit*
*R = Regulatory subunit*

with the purified enzyme (Krebs, 1972; Rosen, Erlichman, and Rubin, 1975). The presence of epinephrine shifts the equilibrium to the right, producing more free catalytic subunits measured as independent or I kinase, together with decreased binding of cyclic AMP after the hormonal treatment. The latter indicates a decrease in the amount of holoenzyme. The inverse would be true with insulin. Hormonal treatment shifts the equilibrium to the left, decreasing the amount of catalytic subunits measured as I or independent kinase. The binding of cyclic AMP was increased after insulin treatment indicating an increase in the amount of holoenzyme. This mechanism was shown to be operative in fat (Soderling, Corbin, and Park, 1973) and in muscle (Walaas, Walaas, and Grønnerød, 1973). Soderling *et al.* demonstrated these effects of hormones on kinase in extracts of adipose tissue pretreated with hormones. The extracts were passed over Sephadex columns to separate the holoenzyme from the catalytic subunits. Epinephrine acted to dissociate the holoenzyme. The action of insulin in this system was only to counteract the action of epinephrine and to maintain or increase the amount of holoenzyme. Walaas and Walaas, on the other hand, demonstrated that insulin alone increased the amount of holoenzyme in diaphragm. Figure 6 is taken from their work and illustrates this change. The effects of epinephrine and of glucagon on the

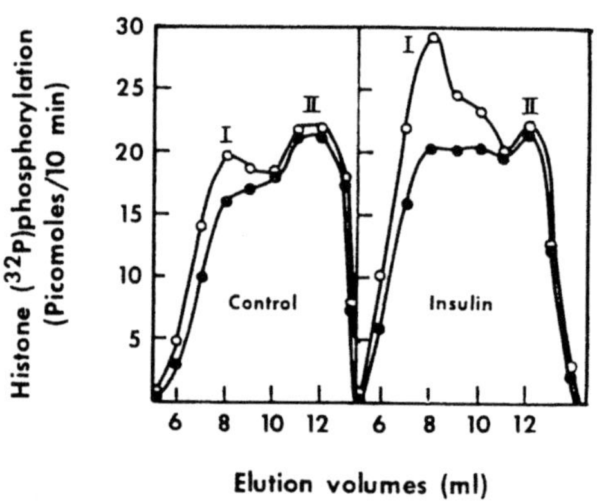


*Figure 6. Gel filtration of protein kinase from extracts of rat diaphragm preincubated with and without insulin. Rat diaphragms were incubated for 10 min with or without insulin. Extracts of rat diaphragms were subjected to gel filtration on Sephadex G-150. Elution was done with 50 mM Tris-Cl pH 7.5+2 mM EDTA containing 0.05 M NaCl. Protein kinase was assayed on 10 μl of each fraction. ●-●: Activity in the absence of cyclic AMP. O-O: Activity in the presence of 1.5 μM cyclic AMP. Peak 1 corresponds with elution volume of RC-complex. Peak 2 corresponds with elution volume of C-subunit.*

kinase appear to be mediated, at least in part, by the increased concentration of cyclic AMP. The effect of insulin may be due to a decreased concentration of cyclic AMP in the presence of epinephrine as in the experiments of Soderling *et al.* Insulin alone may affect the kinase by changing other cellular constituents when basal tissue levels of cyclic AMP are not altered by the hormone. These other constituents would influence the dissociation of the kinase in the presence of a constant level of cyclic AMP or, in other words, affect the sensitivity of the kinase to cyclic AMP. This was, in fact, our proposal as already discussed.

Recently two groups (Avruch, Leone, and Martin, 1976; Benjamin and Singer, 1975) have provided further evidence on this point using isolated adipocytes prelabelled with $^{32}P$ inorganic phosphate. Avruch *et al.* demonstrated that epinephrine enhanced the incorporation of radioactivity into a number of phosphopeptides within the cell. The largest stimulating effect of epinephrine was observed with peptides of molecular weight 123,000 and 69,000 daltons. This effect was further augmented by theophylline suggesting that it is a cyclic AMP response mediated through the protein kinase. In the presence of insulin the incorporation of radioactivity into the 69,000 dalton peptide as well as into a 26,000 dalton peptide was decreased, suggesting a decreased activity of the cyclic AMP protein kinase. However, insulin alone stimulated the incorporation of $^{32}P$ into the 123,000 dalton peptide. Furthermore, insulin augmented the stimulation by epinephrine of $^{32}P$ into the 123,000 dalton peptide. This suggests that insulin can enhance a type of kinase activity, in addition to its action to decrease cyclic AMP-dependent protein kinase. Perhaps insulin is acting to stimulate a special type of kinase which differs from the cyclic AMP-dependent protein kinase. We know from the work of Nishizuka and his colleagues (Kumon, Nishizuka, Yamamura, Nishizuka, 1972) that there are present in tissues several cyclic AMP-independent protein kinases which act selectively on phosphovitin and casein. Cohen and coworkers (Nimmo, Proud, and Cohen, 1976) have purified the phosphovitin kinase and have shown that it acts on glycogen synthase, while Schlender and Reimann (1975) have purified the casein kinase and shown that it also acts on glycogen synthase. It is still too early to evaluate the significance of these enzymes in insulin action, but we must be aware of the fact that they exist and may play a role in addition to that of the cyclic AMP-dependent protein kinase.

There is some evidence, although smaller in amount, that the phosphatase may also be regulated by insulin. This was first shown in 1970 by two workers (Bishop, 1970; Gold, 1970). Gold demonstrated that in a diabetic animal the phosphatase had markedly declined and was restored by insulin treatment. Bishop showed, in short time experiments, that the phosphatase was rapidly activated by insulin and equally rapidly inactivated by glucagon in minutes.

However, since the main argument of this paper is not materially influenced by this additional enzyme parameter altered by hormones, we will not further consider it here, other than to state that the discussion of the kinase is not meant to imply that this is the only site of action of insulin proximal to glycogen synthase.

## THE INSULIN INTERMEDIATE--"X"

Cyclic AMP levels have been measured during insulin stimulated glycogen synthesis and were unchanged from control basal levels (Craig, Rall, and Larner, 1969; Goldberg, Villar-Palasi, Sasko, and Larner, 1967). If levels are first elevated, say by epinephrine or glucagon, they are rapidly reduced by insulin action (Butcher, Sneyd, Park, and Sutherland, 1966). Also, if diaphragms are preincubated with insulin, the subsequent rise in cyclic AMP level brought on by epinephrine is blunted (Craig *et al.*, 1969). This indicates that insulin can act to either lower an already elevated cyclic AMP level or prevent it from being elevated. Under certain circumstances insulin has been shown to elevate cyclic AMP levels (Kono and Barham, 1973). A careful study of the time course of insulin action in liver revealed that the small decrease in cyclic AMP level occurred after the prior activation of glycogen synthase (Villar-Palasi, Goldberg, Bishop, Nuttall, Schlender, and Larner, 1970). The evidence is convincing, therefore, that cyclic AMP is not on the direct route of insulin action. In contrast to the situation in 1970 and before, there is now general agreement on this point (Fain, 1974).

Cyclic GMP has been measured and some workers (Illiano, Tell, Siegel, and Cuatrecassas, 1973) have noted increased levels early during insulin action, while others have noted no effects (Goldberg, Dietz, and O'Toole, 1969; Jarett, Steiner, Smith, and Kipnis, 1972; Larner *et al.*, 1976). Very recently, Fain has made the important observation that the elevation of cyclic GMP levels with insulin is dependent on the presence of $Ca^{2+}$ in the medium, but that the antilipolytic action of insulin occurs without $Ca^{2+}$ in the medium and without the accompanying increase in cyclic GMP concentration. Both the variability of the cyclic GMP response and its dependence on $Ca^{2+}$ argue against placing cyclic GMP on the direct route of insulin action (Fain and Butcher, 1976).

Another very popular current candidate as an intermediate is $Ca^{2+}$. Even though insulin action is known to take place in the absence of external $Ca^{2+}$, it is still possible that insulin can influence $Ca^{2+}$ redistribution within the cell and, thus, place it in the role of a second or third messenger (Clausen, Elbrink, and Dahl-Hanzen, 1975; Kissebah, Hope-Gill, Vydelingum, Tulloch, Clark, and Fraser, 1975). In general, there is evidence that insulin can act to increase $Ca^{2+}$ in the cytoplasm (Clausen *et al.*, 1975). The recent observations of Jarett and his coworkers bear directly on this point (McDonald, Bruns, and Jarett, 1976). These workers demonstrated that insulin altered the pools of $Ca^{2+}$ in fat cell membranes, mitochondria, and microsomes in a differential manner. Mitochondria from insulin-treated cells had 25% less $Ca^{2+}$ as measured by atomic absorption spectroscopy. A similar decrease was observed when EDTA was added at the time of homogenization. No difference was observed when ruthenium red was added. Therefore, insulin did not alter total $Ca^{2+}$ in the mitochondria, but rather insulin decreased a highly stable pool, converting it in effect to a labile pool. On the other hand, insulin-treated cells had a 32% increase in $Ca^{2+}$ content in the microsomes isolated without EDTA, but no effect of insulin was observed on the highly stable pool measured in the microsomes isolated with EDTA. No effect of insulin treatment on the $Ca^{2+}$ of plasma membranes was noted, even though it had been shown in previous experiments that isolated plasma membranes from insulin-treated cells had an increased capacity to bind $^{45}Ca$ (McDonald, Bruns, and Jarett, 1976a). From these observations, they concluded that "A second messenger system other than $Ca^{2+}$ appears necessary to produce the insulin-induced alterations in the calcium content of the two major intracellular organelles, the mitochondria and endoplasmic reticulum." We conclude, therefore, that $Ca^{2+}$ plays an important role in insulin action, but that it may not be on the direct pathway. Similar considerations also apply to other important ions including $K^+$, $Na^+$, and $Mg^{2+}$.

We began to search for "X" by looking for a protein kinase inhibitor generated with insulin action (Larner *et al.*, 1974; Larner *et al.*, 1975). When extracts of control and insulin-treated rabbit hind leg skeletal muscle were deproteinized by TCA or at 100°C, the deproteinized insulin extract demonstrated an increased inhibition of the protein kinase as compared with the control extract (Table 1; Larner *et al.*, 1976). The inhibitor was partially purified on Dowex-1 columns (Table 2). It is eluted in the pH 1.3 fraction with the cyclic GMP. It is, thus, differentiated from an

*Table 1. FRACTIONATION OF INSULIN-MEDIATED PROTEIN KINASE INHIBITOR FROM RABBIT MUSCLE ON DOWEX 1 COLUMNS WITH HCl*

| | Protein kinase activity without cyclic AMP in the presence of | | | Protein kinase activity with added cyclic AMP in the presence of | | |
|---|---|---|---|---|---|---|
| | Control fraction | Insulin fraction | Inhibition | Control fraction | Insulin fraction | Inhibition |
| Fraction | cpm | cpm | % | cpm | cpm | % |
| pass through | 567 | 371 | 34.6 | 2283 | 928 | 59.4 |
| pH 2.7 eluate | 576 | 632 | -- | 3579 | 3873 | -- |
| pH 2.0 eluate | 439 | 384 | 12.5 | 1809 | 1976 | -- |
| pH 1.3 eluate | 279 | 225 | 19.3 | 1288 | 488 | 62.1 |

*Table 2. PRESENCE OF ACID-STABLE PROTEIN KINASE INHIBITOR IN INSULIN-PERFUSED RABBIT SKELETAL MUSCLE*

| | Glycogen Synthase % I form in | | Protein kinase activity with added cyclic AMP in the presence of | | |
|---|---|---|---|---|---|
| Perfusion time min | Control muscle % | Insulin perfused muscle % | Control extract cpm | Insulin extract cpm | Inhibition % |
| 6 | 26.9 | 37.3 | 1029 | 1191 | -- |
| 10 | 26.1 | 37.2 | 603 | 318 | 47.3 |

inhibitor found in the pH 2.0 fraction along with cyclic AMP (Murad and Vaughn, 1969). It is further differentiated from a second protein kinase inhibitor (Walsh, Ashby, Gonzales, Calkins, Fischer, and Krebs, 1961) by the fact that the present insulin inhibitor is TCA soluble, rather than TCA insoluble. In addition, the latter protein kinase inhibitor is absorbed on Dowex I but is not eluted in any of the fractions tested. Finally, it has been reported that levels of the latter protein kinase inhibitor are not changed with insulin action (Wallaas *et al.*, 1973).

While we do not know what the insulin protein kinase inhibitor is, we know that it is not cyclic GMP and that it is extremely labile especially at neutral or alkaline pH. As shown in Figure 7, it disappears irreversibly when it is kept at pH 7 for 30 *min* at 30°C. This lability has made it difficult to purify and to characterize. Needless to say, we are working hard on this problem.

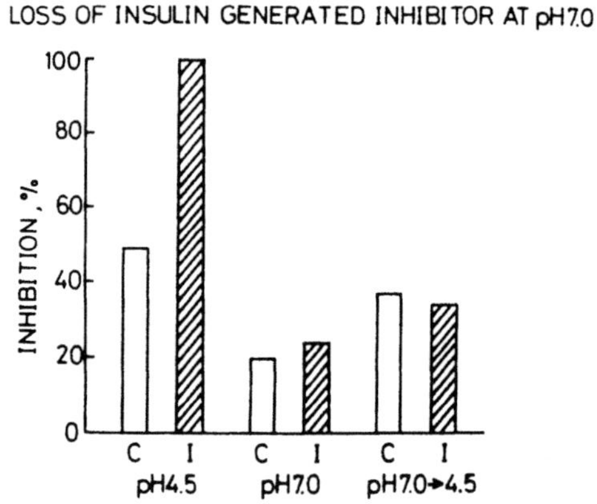


*Figure 7. pH lability of insulin mediated protein kinase inhibitor.*

## MULTIPLE PHOSPHORYLATION OF GLYCOGEN SYNTHASE

In 1971 we first suggested that the subunit of rabbit muscle glycogen synthase D form contained six phosphates as determined by alkali-labile phosphate measurements (Smith, Brown, and Larner, 1971). However, the original $^{32}P$ incorporation data which we had first reported in 1963 (Friedman and Larner, 1963), if analyzed using current values for the specific activity and subunit molecular weight of the enzyme, give a value of five phosphates per 85,000 dalton subunit. On the other hand, others presented evidence that only one phosphate residue per subunit could be introduced into rabbit muscle synthase, as measured by $^{32}P$ incorporation (Soderling, Hickenbottom, Reimann, Hunkeler, Walsh, and Krebs, 1970). Recently, Soderling has reported higher values for the phosphorylation of synthase noting that up to five phosphates could be introduced, depending on the phosphorylating conditions (Soderling, 1975). Others have reported values of about two. Finally, the most recent values from our laboratory have been 3.2 to 3.5 (Roach, Takeda, and Larner, 1976; Takeda and Larner, 1975). With the enzyme from other sources 12 phosphates per subunit have been reported for the rat liver enzyme (Lin and Segal, 1973) and we reported one *mole* of $^{32}P$ incorporated per 20,000 *g* of rat heart enzyme (Larner, Villar-Palasi, and Brown, 1969). If the rat heart enzyme follows other synthases as regards subunit molecular weight, this would correspond to four to five phosphates per subunit. All of the data largely sustain the original evidence for the multiple phosphorylation of the glycogen synthase subunit. Most workers would agree with a stoichiometry of at least two sites per subunit; we argue for a value of between three and six. Furthermore, we now know from limited proteolysis studies that one to two sites are located near the C-terminal "tail" of the molecule and removed by trypsin action, and that one to two are located in the trypsin resistant "core." Figure 8 summarizes these data and demonstrates that the phosphate located in the "tail" as well as in the "core" has an influence on enzyme activity. All the phosphates have an influence on the kinetic properties of the enzyme and none are silent in terms of their effect on the enzyme, as has been claimed by some authors (Nimmo and Cohen, 1974; Nimmo *et al.*, 1976).

Evidently a physiologically significant multiple phosphorylation of glycogen synthase implies the start of a new phase of investigation of the enzyme, namely the search for biological functions for specific phosphorylations and relations between particular sites and particular properties of the enzyme. Little work has been done in this area. As a first step, the quantitative relationship between certain kinetic properties and the average phosphate content of the enzyme has been studied (Roach *et al.*, 1976). The value for the $M_{0.5}$ for glucose-6-P (Figure 9) and the $S_{0.5}$ for UDP-glucose (Figure 10) varied almost exponentially with the average phosphate content of the enzyme in the range 0.27 to 3.5 phosphates

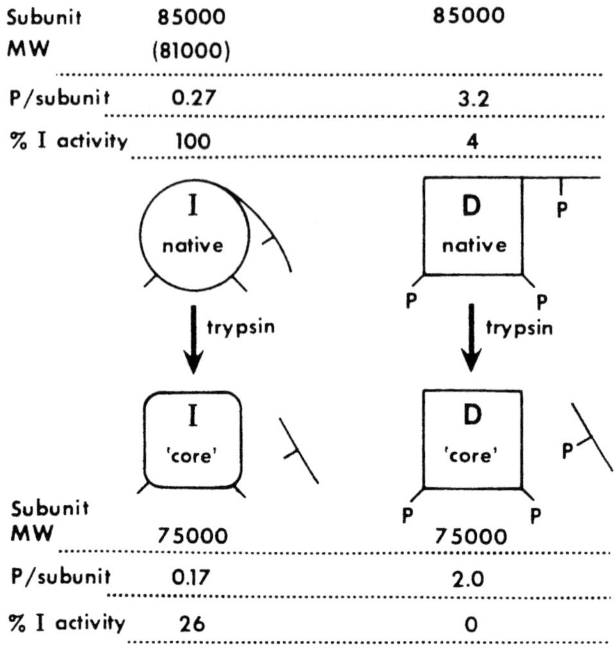


*Figure 8. Effects of limited-tryptic digestion on the properties of rabbit muscle glycogen synthase. Enzyme active without glucose-6-P, is denoted by a circle. Enzyme inactive without glucose-6-P is represented by a square.*

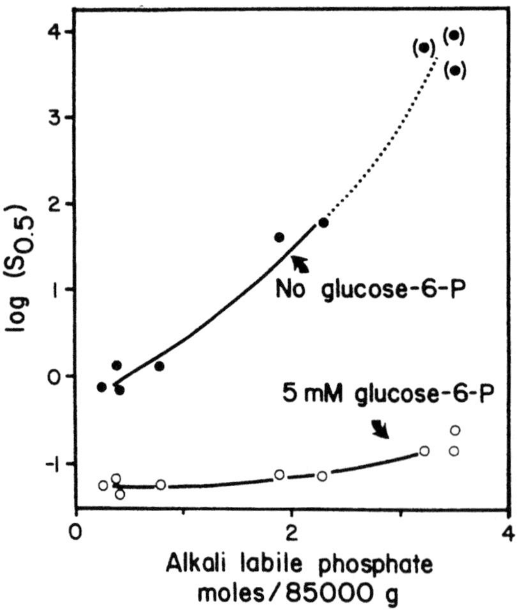


*Figure 9. Relation of glycogen synthase kinetic parameters with covalent phosphorylation state. This figure shows the values of the $S_{0.5}$ (in mM) for UDP-glucose (in the presence O, or in the absence ● of 5 mM glucose-6-P) plotted against the average alkali-labile phosphate content of the rabbit muscle enzyme. The dotted region expresses our uncertainty about the accuracy of the $S_{0.5}$ estimation in the absence of glucose-6-P for the three most phosphorylated enzyme samples.*

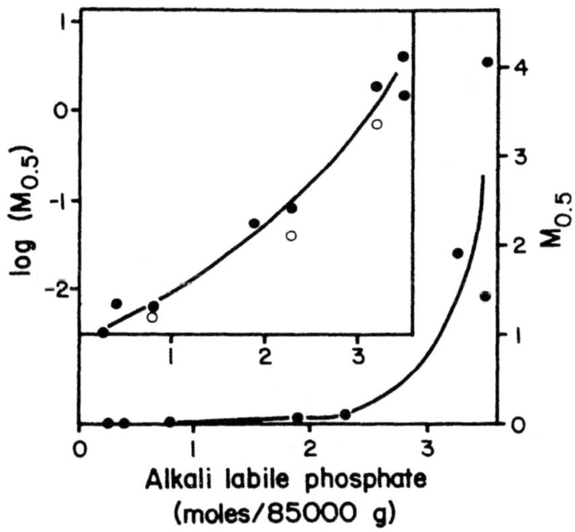


*Figure 10. Relation of glycogen synthase kinetic parameters with covalent phosphorylation state. This figure shows the values of $M_{0.5}$ (in mM) for glucose-6-P (in the presence of 0.2 mM, or 1 mM 0 UDP-glucose) plotted against the average alkali-labile phosphate content of the rabbit muscle enzyme.*

per subunit. With the phosphorylating conditions used, in fact, phosphate could not be introduced into the enzyme without changing the kinetic properties. In further experiments where the average phosphorylation state of synthase was the experimental variable, it was found that, under a variety of conditions, the relation of enzyme activity with phosphate content was nonlinear (Roach and Larner, 1976). The conditions included various glucose-6-P, UDP-glucose, and $Mg^{2+}$ concentrations, as well as the presence or absence of the inhibitors ATP, ADP, AMP, UDP, and $P_i$. An example of this type of experiment is shown in Figure 11. The nonlinearity noted above took the form, in almost all cases, of a steep region at which small changes in phosphate content corresponded to large changes in the expressed activity. Furthermore, the location of the steep region was dependent on the effector concentrations, as is exemplified with the different glucose-6-P concentrations in this figure. Because of the multiple phosphorylation of the glycogen synthase subunit and because of the effects of the multiple phosphorylation on the kinetic parameters of the enzyme, there is a growing inadequacy of the two-state model originally proposed by us (Friedman and Larner, 1963). It has been a model of great utility and importance in research on the control of glycogen metabolism, but at least one point has not been clarified by the model. In general, the magnitude of the changes of % I activity with hormone action is relatively small. For example, in surveying the literature deviations of the % I form from the control

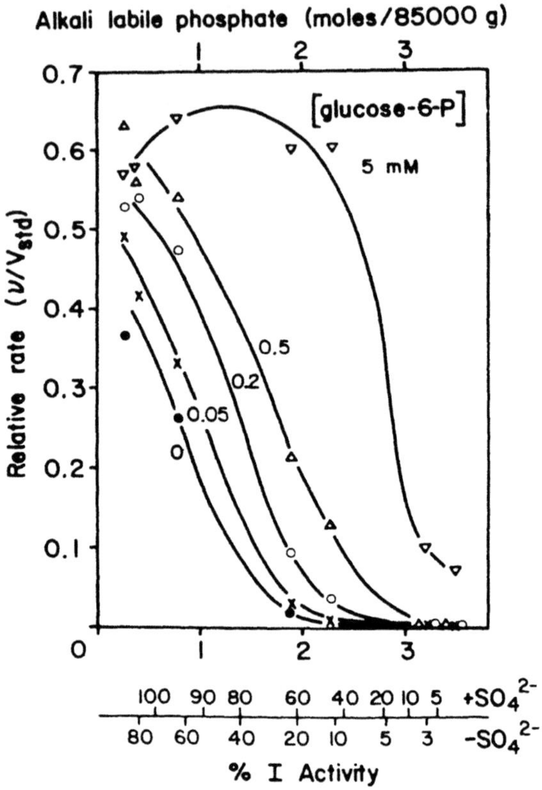


*Figure 11. Reaction rate of glycogen synthase as a function of the enzyme phosphorylation state. Glycogen synthase (rabbit muscle) activity was measured as a function of the alkali-labile phosphate content of the enzyme in the presence of : 0.2 mM UDP-glucose, 0.1 mM UDP, 10 mM $P_i$, 5 mM adenylate mixture (energy charge 0.9) 20 mM $Mg^{2+}$, and glucose-6-P as indicated. The rates have been normalized to the rate under standard conditions of high UDP-glucose and glucose-6-P ($V_{std}$).*

value of 23% to plus or minus 11.5% above or below this value have been determined (Roach and Larner, 1977). On a two-species model of synthase, these rather small changes are difficult to reconcile with an on-off switch type of control usually associated with glycogen synthase. Looking at Figure 11, we can see that is such an experiment is applicable *in vivo*, then even relatively small changes in percent I activity can account for very large alterations in the expressed velocity of glycogen synthase under appropriate conditions of small molecule effectors within the cell. Taking our data of Figure 11 even further, the observed % I activity in tissue extracts would indicate that the enzyme operates

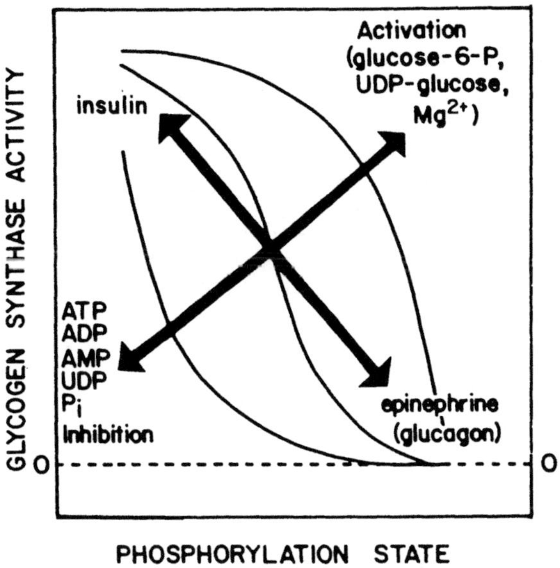


*Figure 12. Control of glycogen synthase activity. Hormone effects cause progression along a solid line (as indicated in the middle curve). Modifiers cause shifts between the solid lines in the sense indicated.*

around a basal phosphorylation state of one to two phosphates per subunit. Noting, also, that the steep portions of the curves are displaced with respect to the phosphorylation axis by changes in effector concentrations, we have suggested that this affords an integration of hormonal control exerted through covalent phosphorylation with local cellular control exerted through effector binding (Roach and Larner, 1976). Thus, as shown in Figure 12, a cell in an energetically favorable state, as reflected by high glucose-6-P levels, will resist the efforts of a hormone stimulus such as epinephrine to switch off the energy storage process. The contrary signal for glycogen synthesis would, on the other hand, be aided by cellular contributions to the regulation of synthase. Similar arguments may be made for a cell in an energy-deprived state. Such a model is clearly somewhat simplistic, but it does provide some refinement of the two species model and allows a rationalization of data such as shown in Figure 11.

Finally, we can compare the control characteristics of a system subject to single phosphorylation and contrast it with a multiple phosphorylation system, and we may compare the regulatory advantages conferred by multiple phosphorylation. We assume that there is no internal symmetry in the subunit so that all phosphorylation sites are distinguishable. The number of distinct species, on the basis of phosphorylation, that are possible for a subunit containing one to six phosphorylation sites is shown in Table 3.

*Table 3. NUMBER OF DISTINCT PHOSPHORYLATED SPECIES OF A SINGLE SUBUNIT AS A FUNCTION OF THE NUMBER OF PHOSPHORYLATION SITES*

| Total number of sites per subunit | Number of distinct species possible: species containing 0 | 1 | 2 | 3 | 4 | 5 | 6 phosphates per subunit | TOTAL |
|---|---|---|---|---|---|---|---|---|
| 1 | 1 | 1 | - | - | - | - | - | 2 |
| 2 | 1 | 2 | 1 | - | - | - | - | 4 |
| 3 | 1 | 3 | 3 | 1 | - | - | - | 8 |
| 4 | 1 | 4 | 6 | 4 | 1 | - | - | 16 |
| 5 | 1 | 5 | 10 | 10 | 5 | 1 | - | 32 |
| 6 | 1 | 6 | 15 | 20 | 15 | 6 | 1 | 64 |

*A single subunit is considered to have from 1 to 6 distinguishable phosphorylation sites. The number of possible monophosphorylated, diphosphorylated, etc. species are related to the total number of sites.*

*Table 4. NUMBER OF DISTINCT SPECIES OF A MULTISUBUNIT ENZYME AS A FUNCTION OF THE NUMBER OF LIGAND BINDING SITES*

| Total number of sites per molecule | Number of distinct species possible: species containing 0 | 1 | 2 | 3 | 4 | 5 | 6 molecules of ligand bound per molecule of enzyme | TOTAL |
|---|---|---|---|---|---|---|---|---|
| 1 | 1 | 1 | - | - | - | - | - | 2 |
| 2 | 1 | 1 | 1 | - | - | - | - | 3 |
| 3 | 1 | 1 | 1 | 1 | - | - | - | 4 |
| 4 | 1 | 1 | 2 | 1 | 1 | - | - | 6 |
| 5 | 1 | 1 | 2 | 2 | 1 | 1 | - | 8 |
| 6 | 1 | 1 | 3 | 4 | 3 | 1 | 1 | 14 |

*Each subunit is assumed to contain a single ligand binding site. Molecules containing from 1 to 6 subunits have been considered, and the number of distinct species, on the basis of ligand binding, computed. A simple cyclic arrangement of subunits has been assumed in all cases, leading to a cyclic symmetry (see text).*

This illustrates the potential limitations of an either I or else D model for synthase. In Table 4 is shown a similar analysis for a multi-subunit enzyme with one ligand binding site per subunit and from one to six subunits per molecule. As soon as multiple subunit systems are considered, the geometrical arrangement of the subunits enters into the analysis and thus, for argument, we have assumed a simple cyclic subunit structure leading to a cyclic symmetry. Thus, each multimer of n subunits has, before ligand binding, an n-fold axis of symmetry. We note that the number of enzyme species distinguishable on the basis of ligand binding is, in general, much less than in the case of the multiple covalent modification in Table 3, the difference arising from the distinguishability of the phosphorylation sites. Thus, the inherent advantage arising from multiple phosphorylation lies in the larger number of distinguishable forms, leading to a finer type of regulatory system capable of assimilating more information bits.

## SUMMARY AND CONCLUSIONS

In summary, we have presented evidence which relates to the action pathway of hormonal control of glycogen metabolism. In the case of insulin, there are changes demonstrable in the cyclic AMP-dependent protein kinase and also in the phosphoprotein phosphatase, under conditions where no direct relationship to either cyclic AMP or cyclic GMP levels are measurable. Therefore, a new unknown intermediate or second messenger system is again proposed. An insulin-generated labile compound(s) which inhibits the protein kinase has been discovered. This may function as an intermediate. Finally, the fact that the glycogen synthase system clearly differs from phosphorylase in its regulation by covalent phosphorylation is discussed. Synthase is now accepted as a multiply phosphorylated subunit, in contrast to phosphorylase which is singly phosphorylated. The inherent theoretical advantages of multiple phosphorylation over single phosphorylation are considered. The advantages of a multistate over a two-state model of enzyme interconversion are mentioned. The importance of the multiple phosphorylations interacting in a nonlinear manner with the control by cellular metabolites is in the explanation of how a small change in covalent phosphorylation signalled by a hormone can be translated in the cell milieu into a much larger change in rate.

## REFERENCES

Avruch, J., Leone, G.R., and Martin, D.B. (1976). Effects of epinephrine and insulin on phosphopeptide metabolism in adipocytes. *J. Biol. Chem.* 251:1511-1515.

Benjamin, W.B. and Singer, I. (1975). Actions of insulin, epinephrine, and dibutyryl cyclic adenosine 5'-monophosphate on fat cell protein phosphorylations, cyclic adenosine 5'-monophosphate dependent and independent mechanisms. *Biochemistry* 14: 3301-3309.

Bishop, J.S. (1970). Inability of insulin to activate liver glycogen transferase D phosphatase in the diabetic pancreatectomized dog. *Biochim. Biophys. Acta* 208:208-218.

Bishop, J.S. and Larner, J. (1967). Rapid activation-inactivation of liver uridine diphosphate glucose-glycogen transferase and phosphorylase by insulin and glucagon *in vivo*. *J. Biol. Chem.* 242:1355-1356.

Butcher, R.W., Sneyd, J.G.T., Park, C.R., and Sutherland, E.W. Jr. (1966). Effect of insulin on adenosine 3',5'-monophosphate in the rat epididymal fat pad. *J. Biol. Chem.* 241:1651-1653.

Clausen, T., Elbrink, J., and Dahl-Hansen, A.G. (1975). The relationship between the transport of glucose and cations across cell membranes in isolated tissues. *Biochim. Biophys. Acta* 375:292-308.

Craig, J.W. and Larner, J. (1964). Influence of epinephrine and insulin on uridine diphosphate glucose-α-glucan transferase and phosphorylase in muscle. *Nature* 202:971-973.

Craig, J.W., Rall, T.W., and Larner, J. (1969). The influence of insulin and epinephrine on adenosine 3',5'-phosphate and glycogen transferase in muscle. *Biochim. Biophys. Acta* 177:213-219.

Fain, J.N. (1974). Mode of action of insulin. pp. 1-23. In: H.V. Rickenberg (Ed.) *Biochemistry of Hormones*, Vol. 8.

Fain, J.N. and Butcher, F.R. (1976). Cyclic guanosine 3',5'-monophosphate and the regulation of lipolysis in rat fat cells. *J. Cyclic Nucl. Res.* 2:71-78.

Friedman, D.L. and Larner, J. (1963). Studies on UDPG-glucan transferase. III. Interconversion of two forms of UDPG-glucan transferase by a phosphorylation-dephosphorylation reaction sequence. *Biochemistry* 2:669-675.

Gold, A.H. (1970). The effect of diabetes and insulin on liver glycogen synthetase activation. *J. Biol. Chem.* 245:903-905.

Goldberg, N.D., Dietz, S.B., and O'Toole, A.G. (1969). Cyclic guanosine 3',5'-monophosphate in mammalian tissues and urine. *J. Biol. Chem.* 244:4458-4466.

Goldberg, N.D., Villar-Palasi, C., Sasko, H., and Larner, J. (1967). Effects of insulin treatment on muscle 3',5'-cyclic adenylate levels *in vivo* and *in vitro*. *Biochim. Biophys. Acta* 148:665-672.

Illiano, T., Tell, G.P.E., Siegel, M.I., and Cuatrecasas, P. (1973). Guanosine 3',5'-cyclic monophosphate and the action of insulin and acetylcholine. *Proc. Nat'l. Acad. Sci.* U.S.A. 70:2443-2447.

Jarett, L., Steiner, A.L., Smith, R.M., and Kipnis, D.M. (1972). The involvement of cyclic AMP in the hormonal regulation of protein synthesis in rat adipocytes. *Endocrinology* 90:1277-1284.

Kissebah, A.H., Hope-Gill, H., Vydelingum, N., Tulloch, B.R., Clark, P.V., and Fraser, T.R. (1975). Mode of insulin action. *Lancet* 1:144-147.

Kono, T. and Barham, F.W. (1973). Effects of insulin on the levels of adenosine 3',5'-monophosphate and lipolysis in isolated rat epididymal fat cells. *J. Biol. Chem.* 248:7417-7426.

Krebs, E.G. (1972). Protein kinases. pp. 99-133. In: B.L. Horecker and E.R. Stadtman (Eds.) *Current Topics in Cellular Regulation*, Vol. 5. Academic Press, New York.

Kumon, A., Nishizuka, K., Yamamura, H., and Nishizuka, Y. (1972). Multiplicity of adenosine 3',5'-monophosphate-dependent protein kinases from rat liver and mode of action of nucleoside 3',5'-monophosphate. *J. Biol. Chem.* 247:3726-3735.

Larner, J. (1972). Insulin and glycogen synthase. *Diabetes* 21: Suppl. 2:428-438.

Larner, J., Huang, L.C., Brooker, G., Murad, F., and Miller, T.B. (1974). Inhibitor of protein kinase formed in insulin treated muscle. *FEBS Lett.* 33:261.

Larner, J., Huang, L.C., Hazen, R., Brooker, G., and Murad, F. (1975). Mechanism of insulin action on glycogenesis. *Diabetes* 24:394.

Larner, J., Takeda, Y., Brewer, H.B., Huang, L.C., Hazen, R., Brooker, G., Murad, F., and Roach, P. (1976). Studies on glycogen synthase and its control by hormones. In: S. Shaltiel (Ed.) *Metabolic Interconversion of Enzymes*. Springer Verlag.

Larner, J., Villar-Palasi, C., and Brown, N.E. (1969). Uridine diphosphate glucose: α-1,4 glucan α-4-glucosyl transferase in heart. Two forms of the enzyme; interconversion reactions and properties. *Biochim. Biophys. Acta* 178:470-479.

Lin, D.C. and Segal, H.L. (1973). Homogeneous glycogen synthetase b from rat liver. *J. Biol. Chem.* 248:7007-7011.

McDonald, J.M., Bruns, D.E., and Jarett, L. (1976a). Ability of insulin to increase calcium binding by adipocyte plasma membranes. *Proc. Nat'l. Acad. Sci.* U.S.A. 73:1542-1546.

McDonald, J.M., Bruns, D.E., and Jarett, L. (1976b). The ability of insulin to alter stable calcium pools of isolated adipocyte subcellular fractions. *Biochem. Biophys. Res. Commun.* 71:114-121.

Miller, T.B. Jr. and Larner, J. (1973). Mechanism of control of hepatic glycogenesis by insulin. *J. Biol. Chem.* 248:3483-3488.

Miller, T.B. Jr. and Larner, J. (1972). Anti-insulin actions of a bovine pituitary diabetogenic peptide on glycogen synthesis. *Proc. Nat'l. Acad. Sci.* U.S.A. 69:2774-2777.

Murad, F., Rall, T., and Vaughan, M. (1969). Conditions for the formation, purification and assay of an inhibitor of adenosine 3',5'-monophosphate. *Biochim. Biophys. Acta* 192:430-445.

Nimmo, H.B. and Cohen, P. (1974). Glycogen synthetase kinase 2 ($GSK^2$). The identification of a new protein kinase in skeletal muscle. *FEBS Lett.* 47:162-166.

Nimmo, H.G., Proud, C.G., and Cohen, P. (1976). The phosphorylation of rabbit skeletal muscle glycogen synthase by glycogen synthase kinase 2 and adenosine 3',5'-monophosphate dependent protein kinase. *Eur. J. Biochem.* 68:31-44.

Nuttall, F.Q. and Larner, J. (1971). Studies on glycogen synthetase interconverting enzymes in *in vitro* perfused compared to non-perfused rat hearts. *Biochim. Biophys. Acta* 230:560-570.

Roach, P.J. and Larner, J. (1976). Rabbit skeletal muscle glycogen synthase. II. Enzyme phosphorylation state and effector concentrations as interacting control parameters. *J. Biol. Chem.* 251:1920-1925.

Roach, P.J. and Larner, J. (1977). Covalent phosphorylation in the regulation of glycogen synthase activity. *Mol. Cell. Biochem.* 15:179-200.

Roach, P.J., Takeda, Y., and Larner, J. (1976). Rabbit muscle glycogen synthase. I. Relationship between phosphorylation state and kinetic properties. *J. Biol. Chem.* 251:1913-1919.

Rosen, O.M., Erlichman, J., and Rubin, C.S. (1975). Molecular structure and characterization of bovine heart protein kinase. pp. 253-263. In: G.I. Drummond, P. Greengard, and G.A. Robison (Eds.) *Advances in Cyclic Nucleotides Research*, Vol. 5. Raven Press, New York.

Schlender, K.K. and Reimann, E.M. (1975). Isolation of a glycogen synthase I kinase that is independent of adenosine 3',5'-monophosphate. *Proc. Nat'l. Acad. Sci.* U.S.A. 72:2197-2201.

Shen, L.C., Villar-Palasi, C., and Larner, J. (1970). Hormonal alteration of protein kinase sensitivity to cyclic AMP. *Phys. Chem. and Phys.* 2:536-544.

Smith, C.H., Brown, N.E., and Larner, J. (1971). Molecular characteristics of the totally dependent and independent forms of glycogen synthase of rabbit skeletal muscle II. *Biochim. Biophys. Acta* 242:81-88.

Soderling, T.R. (1975). Regulation of glycogen synthetase. *J. Biol. Chem.* 250:5407-5412.

Soderling, T.R., Corbin, J.D., and Park, C.R. (1973). Regulation of adenosine 3',5'-monophosphate-dependent protein kinase II. Hormonal regulation of the adipose tissue enzyme. *J. Biol. Chem.* 248:1822-1829.

Soderling, T.R., Hickenbottom, J.P., Reimann, E.M., Hunkeler, F.L., Walsh, D.H., and Krebs, E.G. (1970). Inactivation of glycogen synthetase and activation of phosphorylase kinase by muscle adenosine 3',5'-monophosphate-dependent protein kinases. *J. Biol. Chem.* 245:6317-6328.

Takeda, Y. and Larner, J. (1975). Structural studies on rabbit muscle glycogen synthase. II. Limited proteolysis. *J. Biol. Chem.* 250:3951-3956.

Villar-Palasi, C., Goldberg, N.D., Bishop, J.S., Nuttall, F.W., Schlender, K.K., and Larner, J. (1970). Hormonal control of glycogen synthetase interconversions. pp. 161-180. In: J.J. Blum (Ed.) *Soc. Gen'l. Phys. Symposium on Biogenic Amines.* Prentice Hall, New York.

Villar-Palasi, C. and Wenger, J.I. (1967). *In vivo* effect of insulin on muscle glycogen synthetase. Identification of the action pathway. *Fed. Proc.* 26:563.

Walaas, O., Walaas, E., and Grønnerød, O. (1973). Hormonal regulation of cyclic-AMP-dependent protein kinase of rat diaphragm by epinephrine and insulin. *Eur. J. Biochem.* 40:465-477.

Walaas, O., Walaas, E., and Grønnerød, O. (1972). Effect of insulin and epinephrine in cyclic AMP-dependent protein kinase in rat diaphragm. *Israel J. Med. Sci.* 8:353-357.

Walsh, D.A., Ashby, C.D., Gonzales, C., Calkins, D., Fischer, E.H., and Krebs, E.G. (1971). Purification and characterization of a protein inhibitor of adenosine 3',5'-monophosphate-dependent protein kinase. *J. Biol. Chem.* 246:1977-1985.

# HORMONAL CONTROL OF GLUCONEOGENESIS

J.H. Exton[1]

Department of Physiology
Vanderbilt University
School of Medicine
Nashville, TN 37232

## OUTLINE



---

[1]Investigator and Director, Howard Hughes Medical Institute Laboratories for the Studies of Metabolic Disorders. Unpublished studies supported by grants 5 P01 AM 07462, 1 R01 AM 18660, and 1 P17 AM 17026 from the National Institutes of Health, U.S. Public Health Service.

## INTRODUCTION

Gluconeogenesis is the process by which glucose and glycogen are synthesized in the animal body from noncarbohydrate precursors. The liver and the kidney are the two organs which carry out gluconeogenesis and gluconeogenic substrates include lactate, pyruvate, glycerol, and the glucogenic amino acids. Although all the natural amino acids except leucine and lysine are potentially glucogenic by virtue of the fact that they yield pyruvate, oxalacetate, α-ketoglutarate, succinyl-CoA, or fumarate during their catabolism, studies in the perfused liver indicate that only alanine, serine, proline, threonine, glutamine, asparagine, glutamate, aspartate, and arginine yield significant amounts of carbohydrate (Ross, Hems, and Krebs, 1967).

Gluconeogenesis is under complex control by hormones such as insulin, glucagon, catecholamines, and glucocorticoids and changes in gluconeogenesis play an integral part in the adaptations of the organism during starvation, exercise, and the neonatal and suckling periods. In addition, changes in gluconeogenesis are important in certain pathological conditions such as diabetes and adrenal insufficiency.

Hormonal regulation of gluconeogenesis may be exerted directly on the liver or indirectly by control of the output of gluconeogenic substrates from peripheral tissues such as muscle and fat. Frequently, hormones exert effects at both sites concurrently. The effects may be exerted rapidly through covalent or allosteric modification of enzymes or may occur more slowly through changes in enzyme or protein synthesis or degradation.

In this article, the hormonal control of gluconeogenesis will be reviewed with emphasis on the *in vivo* situation and on the biochemical mechanisms involved. For a detailed description of the gluconeogenic pathway and of the properties of key enzymes on which control is exerted, the reader is referred to Exton (1972), Hanson and Garber (1972), Newsholme and Gevers (1967), and Scrutton and Utter (1968).

## REGULATION OF GLUCONEOGENIC SUBSTRATE SUPPLY

### Role of Insulin

Studies in eviscerated or functionally hepatectomized animals have indicated that skeletal muscle releases amino acids, mainly glutamine and alanine, during the postabsorptive state and that insulin produces a striking reduction in this release (for references see Blackshear, Holloway, and Alberti, 1974; Pozefsky, Felig, Tobin, Soeldner, and Cahill, 1969). Measurements of amino acid balance across the forearm tissues of man have yielded similar results

(Pozefsky *et al.*, 1969). Raising the insulin level in forearm deep veins from 12 μU/mℓ to a high, but still physiological level (157 μU/mℓ) reduces α-amino N release by 74% with consistent declines in leucine, isoleucine, threonine, glycine, α-aminobutyric acid, phenylalanine, and tyrosine. Since the latter two amino acids are neither synthesized nor catabolized by skeletal muscle, these findings are consistent with stimulation of protein synthesis and/or inhibition of protein breakdown by insulin.

Studies utilizing perfused heart, perfused hemicorpus, and incubated diaphragm preparations have similarly shown that insulin can inhibit net amino acid release. Results from the studies of Jefferson, Li, and Rannels (1977), using the perfused rat hemicorpus, are presented in Figure 1. It is seen that insulin causes significant decreases in the output of all amino acids measured except aspartate and glutamate. The effect is minimal, or absent, during the first hour of perfusion due, perhaps, to the persistence of the action of endogenous insulin. By measuring changes in the incorporation of [$^{14}$C]phenylalanine into protein under conditions in which the intracellular specific radioactivity of the amino acid was kept constant or was determined, Jefferson, Koehler,

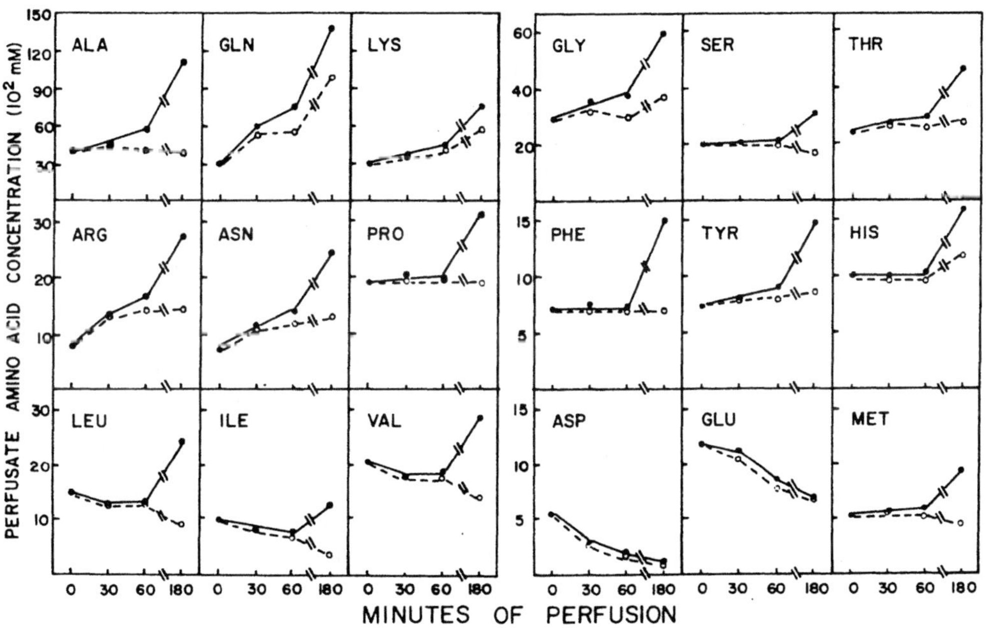


*Figure 1. Effects of insulin (25 mU/mℓ) on the release of amino acids from the perfused rat hemicorpus. Solid lines, control perfusions; broken lines, insulin perfusion. Data are from Jefferson* et al. *(1977).*

and Morgan (1972) and Jefferson *et al.* (1977) have obtained evidence that the effects of insulin are due to concurrent stimulation of protein synthesis and inhibition of protein breakdown. Using a similar approach with [$^{14}$C]tyrosine and by measuring tyrosine release in the presence of cycloheximide, Fulks, Li, and Goldberg (1975) have reached similar conclusions in rat diaphragm. The effect of protein synthesis appears to involve, in part at least, increased ribosomal aggregation, *i.e.*, acceleration of steps involved in the initiation of peptide chains (Jefferson *et al.*, 1972; Jefferson *et al.*, 1977; Morgan, Jefferson, Wolpert, and Rannels, 1971). The action on protein degradation may involve reduced free lysosomal protease activity (Jefferson *et al.*, 1977).

Insulin has also been reported to stimulate the transport of alanine and the model amino acid α-aminoisobutyrate into several muscle preparations (for references see Narahara and Holloszy, 1974). However, since studies in the perfused hemicorpus show that the tissue levels of most amino acids decline with insulin in about the same proportion as their release into the medium decreases (Jefferson *et al.*, 1977), it appears that any effects of the hormone on amino acid transport are of minor importance compared with the changes due to alterations in protein metabolism.

Infusion of physiological amounts of insulin into the forearm of postabsorptive man inhibits the net release of free fatty acids, presumably from adipose tissue (Pozefsky *et al.*, 1969). The hormone also markedly reduces the accumulation of glycerol in the bloodstream of functionally hepatectomized rats (Blackshear *et al.*, 1974). These changes reflect the antilipolytic action of insulin which has been well demonstrated in adipose tissue *in vitro* and is due, in part at least, to a reduction in cyclic AMP levels with consequent decreased activity of cyclic AMP-dependent protein kinase and hormone-sensitive lipase (Soderling, Corbin, and Park, 1973).

In addition to its ability to markedly inhibit the inflow of glycerol and glucogenic amino acids into the bloodstream, insulin also reduces the peripheral release of lactate, another major glucose precursor. This effect is seen in functionally hepatectomized rats (Blackshear *et al.*, 1974). However, it is not evident in *in vitro* preparations such as perfused rat hindquarters (Caldwell, Lacy, and Exton, 1976; Ruderman, Houghton, and Hems, 1971; Theinhaus, Tharandt, Zais, and Staib, 1975), suggesting that it requires factors missing from the perfusion system or is exerted indirectly or on tissues other than muscle. The inability of insulin to increase lactate (and alanine) production in muscle preparations in the face of marked acceleration of glucose uptake indicates that the hormone also affects glucose metabolism beyond glucose transport. Probable sites are glycogen synthetase and pyruvate dehydrogenase.

## Role of Glucocorticoids

The classic studies of Long, Katzin, and Fry (1940) and Smith and Long (1967) in adrenalectomized rats indicate that glucocorticoids exert part of their control of gluconeogenesis by stimulating the net release of amino acids from peripheral tissues (for additional references see Steele, 1975). Recent experiments by Caldwell *et al.* (1976) and Theinhaus *et al.* (1975) utilizing perfused hindquarters from fasted rats have also shown that adrenalectomy causes a marked reduction in the release of glucogenic and branched chain amino acids from skeletal muscle and that glucocorticoids reverse this. As illustrated in Figure 2, the changes in amino acid release with adrenalectomy are accompanied by decreases or no alterations in the intracellular concentrations of the affected amino acids indicating that they result from decreases in net amino acid production. Since in these experiments neither steroid depletion nor repletion significantly altered the release of phenylalanine or tyrosine (Figure 2) and had minimal or no effects on the tissue levels of these amino acids, the amino acid changes cannot be attributed to alterations in net protein catabolism. In perfusion studies of longer duration, Rannels, Li, and Jefferson (1976) have observed increased protein synthesis in hemicorpora from fasted adrenalectomized rats. These workers also observed that cortisol treatment depressed protein synthesis, but in contrast to earlier suggestions, found no changes in protein degradation with adrenalectomy or steroid treatment. Furthermore, comparison of the amino acid changes observed in perfusion studies with those attributable to alterations in protein synthesis indicates clearly that other biochemical changes are mainly responsible for the marked alterations in production of glycogenic and branched chain amino acids induced by glucocorticoids. The nature of these changes remains to be elucidated.

Glucocorticoids also influence gluconeogenesis by virtue of their effects on the actions of catecholamines and the sympathetic nervous system on glycerol output from adipose tissue and lactate release from muscle. In tissues from adrenalectomized animals, the lipolytic and glycogenolytic responses to epinephrine and other hormones acting through cyclic AMP are reduced (for references see Exton, Friedmann, Wong, Brineaux, Corbin, and Park, 1972; Steele, 1975). The impairments are attributable to diminished responsiveness to cyclic AMP rather than to reduced cyclic AMP generation (Exton *et al.*, 1972; Miller, Exton, and Park 1971). The data of Miller *et al.* (1971) and other unpublished results indicate that there is no defect in the activation of protein kinase or phosphorylase kinase in steroid-deficient tissues, indicating that glucocorticoids act beyond these enzymes. Exton *et al.* (1972) have suggested that the action of these steroids may involve the maintenance of normal intracellular concentrations of $Ca^{++}$ and other ions thus permitting normal responses of enzyme systems to

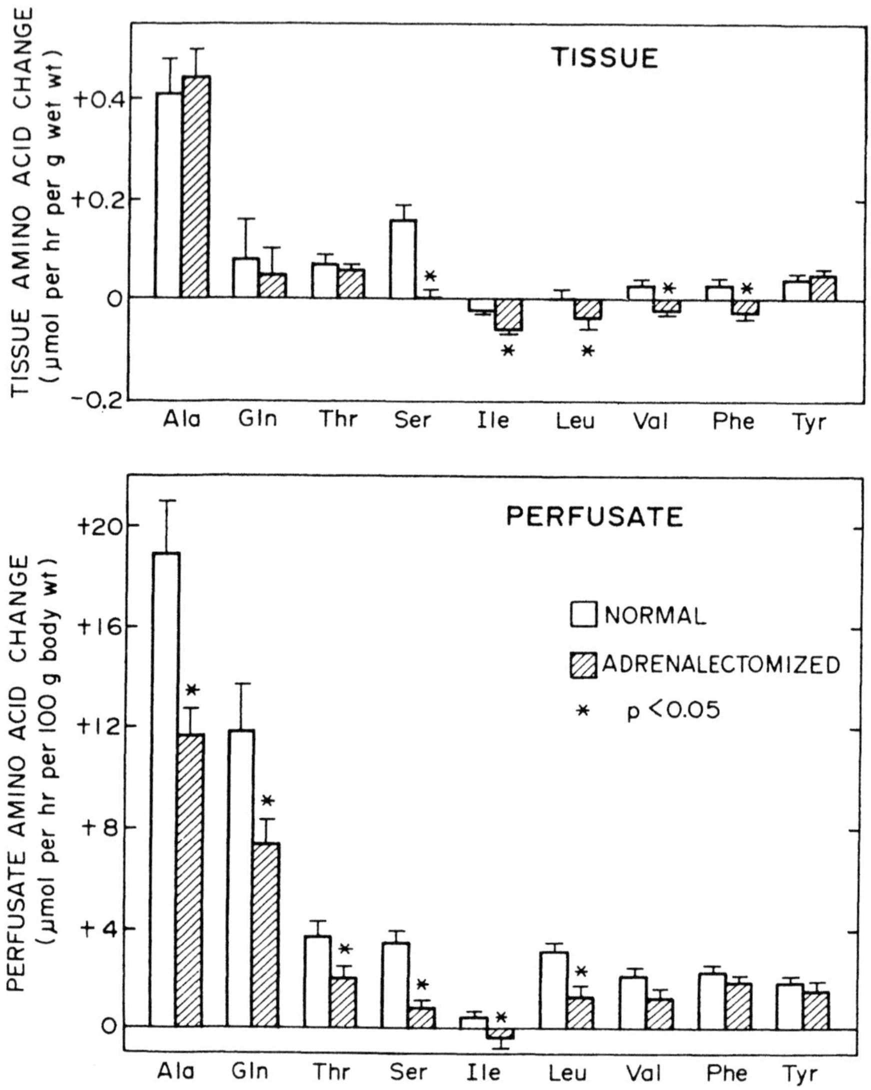


*Figure 2.* *Effects of adrenalectomy on amino acid changes in perfusate and tissue during perfusion of hindquarters from fasted rats. The lower panel shows the perfusate amino acid changes and the upper panel shows the changes in free amino acids in the gastrocnemii. Data are from unpublished studies by Caldwell et al. (1976).*

cyclic AMP. This theory has not been proven.

Glucocorticoids probably also modulate gluconeogenesis through their ability to decrease the sensitivity of peripheral tissues to insulin. This hormonal interaction has been explored, as it relates to glucose transport and phosphorylation in the heart by Morgan, Regen, Henderson, Sawyer, and Park (1961), but it has not been examined in skeletal muscle *in vitro* or in relation to the effects of insulin on protein metabolism. However, it seems very likely that glucocorticoids can counteract the inhibitory effect of insulin on amino acid release from muscle tissue. For example, Manchester, Randle, and Young (1959) found that adrenalectomy increased and cortisol treatment decreased the stimulatory effect of insulin on [$^{14}$C]glycine incorporation into diaphragm protein.

The interaction between glucocorticoids and insulin in the regulation of glycerol output has been little studied. Physiological levels of glucocorticoids by themselves do not stimulate glycerol release from adipose tissue, although they may stimulate free fatty acid release in the presence of glucose (Fain and Czech, 1975). It would seem unlikely that these steroids would antagonize the antilipolytic action of insulin since they do not alter insulin binding or cyclic AMP levels in fat cells *in vitro* (Fain and Czech, 1975; Olefsky, 1975).

## Role of the Sympathetic Nervous System

Catecholamines enhance glycerol and lactate output from peripheral tissues by interacting with β-adrenergic receptors in adipose tissue and muscle. This interaction leads to activation of adenylate cyclase, intracellular accumulation of cyclic AMP and activation of cyclic AMP-dependent protein kinase. In fat, protein kinase phosphorylates and activates the lipase which converts triglycerides to diglycerides and further hydrolysis of diglycerides leads to the release of glycerol. In muscle, the enzyme phosphorylates and activates phosphorylase kinase and phosphorylates and inactivates the active or I form of glycogen synthase. These changes lead to increased glycogenolysis, through phosphorylase activation, and decreased glycogen synthesis. As a result, glycolytic flux increases and lactate production ensues (for references see Soderling and Park, 1974; Williamson, 1975). Epinephrine may also enhance muscle lactate output through inhibitory effects on pyruvate dehydrogenase, which are apparently not mediated by cyclic AMP directly or by an increase in intracellular $Ca^{++}$ leading to increased phosphorylase kinase activity. The latter effect may involve cyclic AMP and phosphorylation of membrane proteins or may occur concurrently with adenylate cyclase activation (Williamson, 1975).

Surprisingly, catecholamines acting through β-adrenergic receptors inhibit the release of alanine and other amino acids from

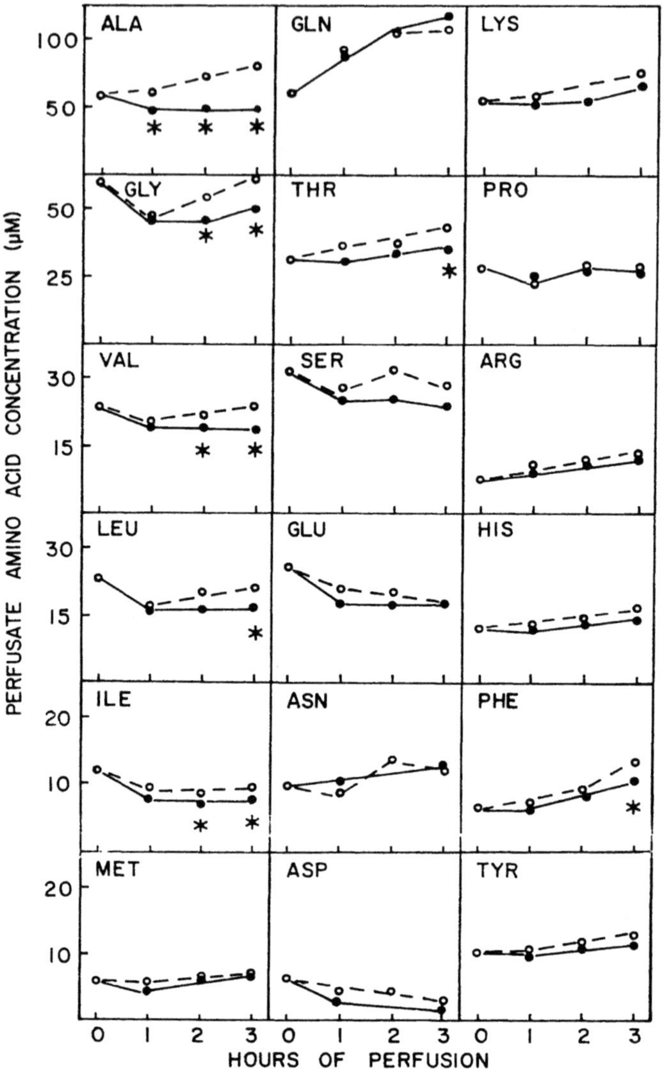


*Figure 3. Effect of $10^{-6}$ M isoproterenol on the release of amino acids from the perfused rat hemicorpus. Broken lines, control perfusions; solid lines, isoproterenol perfusions. Data are from Li and Jefferson (1976).*

perfused muscle preparations (Garber, Karl, and Kipnis, 1976; Li and Jefferson, 1976). This is shown in Figure 3 from the studies of Li and Jefferson (1976). It would have been thought that these hormones would have increased protein catabolism and promoted alanine output through enhancement of glycolysis. Although epinephrine increases the uptake of nonutilizable amino acids by cardiac muscle, this does not occur in skeletal muscle (Sanders and Riggs,

1967). An action on membrane transport is also excluded as a major mechanism by measurements of tissue amino acid levels (Garber *et al.*, 1976). Li and Jefferson (1976) have obtained evidence that the β-adrenergic agonist isoproterenol inhibits protein catabolism and suggest that it may also increase alanine utilization.

## Role of Other Hormones

Although growth hormone promotes protein anabolism in most tissues, its role in the regulation of gluconeogenesis from amino acids and other substrates has received little attention. Many studies have shown that it promotes the transport of amino acids and the synthesis of protein in skeletal muscle (for references see Kostyo and Nutting, 1974). The stimulation of protein synthesis occurs slowly and appears to involve increased ability of ribosomes to carry out peptide chain initiation and elongation. Obviously, these effects of growth hormone on muscle protein metabolism would result in decreased glucoenogenesis from amino acids *in vivo*.

Growth hormone also increases fatty acid and glycerol release *in vivo* and *in vitro* by enhancing triglyceride breakdown and depressing fatty acid reesterification (Gerich, Lorenzi, Bier, Tsalikian, Schneider, Karam, and Forsham, 1976; Goodman and Schwartz, 1974). The effect is slow and apparently involves protein synthesis. In studies in juvenile-onset diabetics treated with somatostatin, Gerich *et al.* (1976) have shown that physiological levels of growth hormone can augment lipolysis, but that this action is not seen in the presence of physiological levels of insulin. As noted by Cahill (1970) and others, glycerol is a relatively minor gluconeogenic substrate.

Glucagon exerts potent effects on hepatic metabolism, but has minimal effects on peripheral tissues. It promotes glycerol and fatty acid release at apparently physiological concentrations in humans provided insulin is absent (Gerich *et al.*, 1976; Liljenquist, Bomboy, Lewis, Sinclair-Smith, Felts, Lacy, Crofford, and Liddle, 1974). It also causes transient falls in the plasma levels of the glucogenic amino acids glutamine, alanine, glycine, and slower rises in the branched chain amino acids in fasting subjects (Aoki, Muller, Brennan, and Cahill, 1974). These changes undoubtedly reflect, in part, increased gluconeogenesis but the elevations in leucine, valine, and isoleucine suggest that their oxidation in skeletal muscle may have declined with consequent reduction in alanine release. These branched chain amino acids are only slowly metabolized by liver, but are readily utilized by skeletal muscle where they donate their amino groups to pyruvate to form alanine (Odessey, Khairallah, and Goldberg, 1974). Since insulin levels declined during glucagon infusion in the studies of Aoki *et al.* (1974), it is unclear to what extent the observed changes were due to glucagon *per se*. In experiments utilizing perfused rat hindquarters, Ruderman and Berger (1974) were unable to observe effects of glucagon on the release of alanine or glutamine, but data on other amino acids are lacking.

## REGULATION OF HEPATIC GLUCONEOGENESIS AND GLYCOLYSIS

### Role of Glucagon

Physiological levels of glucagon rapidly activate gluconeogenesis from lactate, pyruvate, alanine, and glycerol in the perfused rat liver and isolated hepatocytes (Claus, Pilkis, and Park, 1975; Exton and Park, 1968; Harper and Exton, 1976; Hutson, Brumley, Assimacopoulos, Harper, and Exton, 1976; Mallette, Exton, and Park, 1969a; Pilkis, Riou, and Claus, 1976). Since these effects can be mimicked by exogenous cyclic AMP and are preceded by a rise in the nucleotide and activation of cyclic AMP-dependent protein kinase (Cherrington and Exton, 1977), it is considered that they are mediated by this nucleotide and enzyme. However, efforts to link cyclic AMP or protein kinase to the phosphorylation or activations of specific gluconeogenic enzymes *in vivo* or *in vitro* have not been successful and the precise site(s) of action of glucagon on this process remain unknown.

Measurements of tissue levels of metabolic intermediates and isotope studies in liver preparations have identified two sequences in the glucogenic pathway as probable sites of glucagon or cyclic AMP action (Clark, Kneer, Bosch, and Lardy, 1974; Exton and Park, 1969; Ui, Claus, Exton, and Park, 1973; Ui, Exton, and Park, 1973; Veneziale, 1971). These are between pyruvate and P-pyruvate and between fructose-1, 6-di-P and glucose-6-P. In addition, glucagon can stimulate the transport of alanine and other amino acids into liver cells (Mallette *et al.*, 1969b). Possible mechanisms of action of glucagon on P-pyruvate synthesis are described in more detail by Exton, Ui, Lewis, and Park (1971) and Parrilla, Jimenez, and Ayoso-Parilla (1976). More recently, indirect evidence has been presented for an effect of glucagon on flux through fructose 1,6-diphosphatase (Clark *et al.*, 1974).

Glucagon also rapidly inhibits hepatic glycolysis. This is illustrated in Figure 4 from Pilkis *et al.* (1976) which shows that hormone reduces the conversion of $[^{14}C]$dihydroxyacetone to $[^{14}C]$ lactate in hepatocytes from fed rats. Similar findings have been made using $[^{14}C]$glycerol. It is probable that the effect is due to the inhibitory action of glucagon on L-type rat liver pyruvate kinase recently observed by Clair, Cimbala, Foster, and Morgan (1976); Feliu, Hue, and Hers (1976); and Pilkis *et al.* (1976). Several papers from Engstrom's group (Ekman, Dahlqvist, Humble, and Engstrom, 1976; Lungstrom, Hjelmquist, and Engstrom, 1974; Titanji, Zetterqvist, and Engstrom, 1976) have reported that this enzyme can be phosphorylated by cyclic AMP-dependent protein kin-

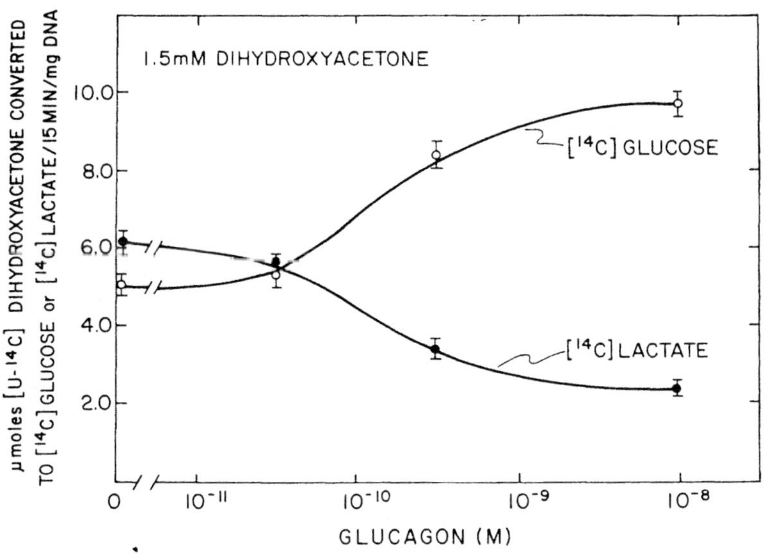


*Figure 4. Dose response curves for the effects of glucagon on [$^{14}C$] glucose and [$^{14}C$]lactate formation from [$^{14}$] dihydroxyacetone in isolated rat liver cells. From Pilkis et al. (1976).*

ase with loss of activity when assayed at low substrate concentrations. The phosphorylated form of the enzyme can also be dephosphorylated by a histone phosphatase from rat liver with restoration of full activity (Titanji *et al.*, 1976). Thus, these studies provide a molecular mechanism for the inhibitory effect of glucagon on hepatic glycolysis. Obviously, this effect could contribute to the enhancement of gluconeogenesis since recycling of substrate through pyruvate kinase is substantial in liver (Friedman, Goodman, Saunders, Kostos, and Weinhouse, 1971). However, Felui *et al.* (1976) noted some discrepancies between the effects of glucagon on the enzyme and on gluconeogenesis suggesting that the hormone acts through other mechanisms.

Clark *et al.* (1974) have also obtained evidence for inhibition of P-fructokinase by glucagon in hepatocytes. Brand and Soling (1975) have reported that this enzyme can be phosphorylated by a protein kinase. However, the reaction does not require cyclic AMP and is associated with increased rather than reduced activity. Thus no explanation for the postulated effect of glucagon on P-fructokinase currently exists.

Attempts to demonstrate the stimulation of gluconeogenesis by glucagon *in vivo* have not always shown clear effects. This is because glucagon administration has usually caused insulin release. Recently, Chaisson and coworkers (Chaisson, Cook, Liljenquist, and

Lacy, 1974; Chiasson, Liljenquist, Sinclair-Smith, and Lacy, 1975) have shown that near-physiological levels of glucagon stimulate gluconeogenesis from alanine in dog and man. Since glucagon did not increase alanine extraction by the liver, the authors concluded that the hormone acted by shunting alanine metabolism within the liver towards gluconeogenesis (Chiasson *et al.*, 1975). When endogenous glucagon is suppressed by somatostatin administration *in vivo* and normal insulin levels are maintained by infusion, the effects of a physiological deficiency of glucagon on hepatic metabolism are more discernible. As shown in Figure 5 (second panel) from the studies of Jennings, Cherrington, Liljenquist, Keller, Lacy, and Chiasson (1977), the conversion of [$^{14}$C]alanine into [$^{14}$C]glucose is decreased under these conditions. When physiological levels of glucagon are restored (fourth panel), gluconeogenesis returns to normal, *i.e.*, to the rate seen before somatostatin or in control experiments.

## Role of the Sympathetic Nervous System

Catecholamines rapidly activate hepatic gluconeogenesis (Exton and Park, 1968). Initially, this was ascribed to their inability to increase intracellular cyclic AMP. However, more recent studies with adrenergic blockers and specific α- and β-adrenergic agonists indicate that the effect is predominantly mediated by α-receptors and is largely unrelated to the increase in cyclic AMP (Exton and Harper, 1975; Hutson *et al.*, 1976; Kneer, Bosch, Clark, and Lardy, 1974; Tolbert, Butcher, and Fain, 1973). This is illustrated in Figure 6 which shows that the β-adrenergic blocker propranolol has relatively little inhibitory action on the stimulation of gluconeogenesis or glycogenolysis by epinephrine even though it virtually abolishes the increase in cyclic AMP. On the other hand, the α-blockers phenoxybenzamine and phentolamine significantly inhibit the effects of the catecholamine on glucose synthesis without affecting cyclic AMP.

*Figure 5. Effects of somatostatin infusion with or without replacement of insulin and/or glucagon on splanchnic synthesis of [$^{14}$C]glucose from [$^{14}$C]alanine in postabsorptive dogs in vivo. Somatostatin (1 μg/kg/min) was infused peripherally and insulin and glucagon were infused intraportally to maintain physiological levels. Glucagon and insulin immunoreactivities were measured in arterial blood. Splanchnic conversion of alanine to glucose was calculated from the hepatic vein-arterial differences in [$^{14}$C]glucose, the splanchnic plasma flow and the arterial alanine or splanchnic uptake of alanine during the course of the experiments. Four experiments were performed in each situation. Infusion of saline in place of somatostatin produced no changes in any parameter measured. Data are from Jennings et al. (1977).*

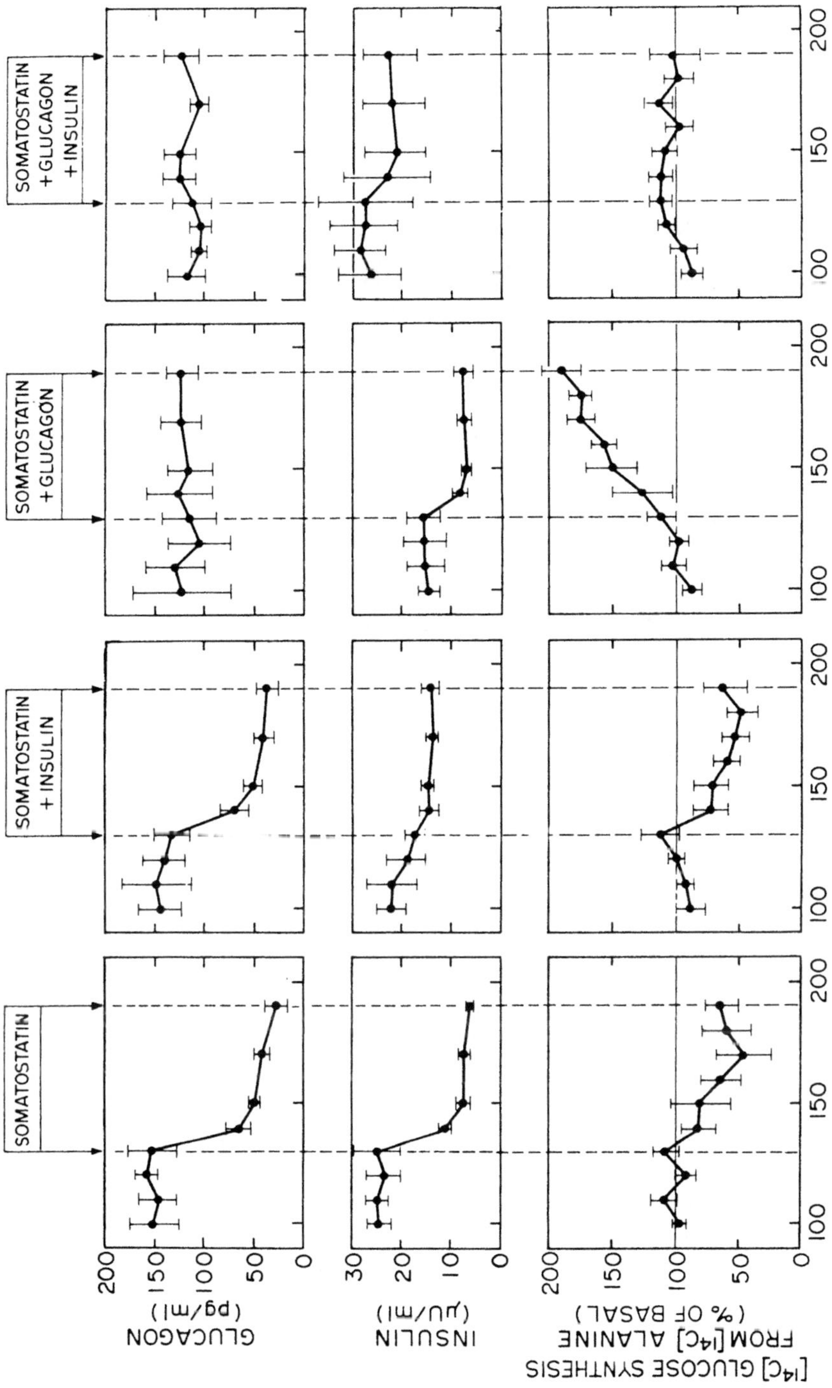
SOMATOSTATIN
SOMATOSTATIN + INSULIN
SOMATOSTATIN + GLUCAGON
SOMATOSTATIN + GLUCAGON + INSULIN
GLUCAGON (pg/ml)
INSULIN (μU/ml)
[14C] GLUCOSE SYNTHESIS FROM [14C] ALANINE (% OF BASAL)
MINUTES

*Figure 5*

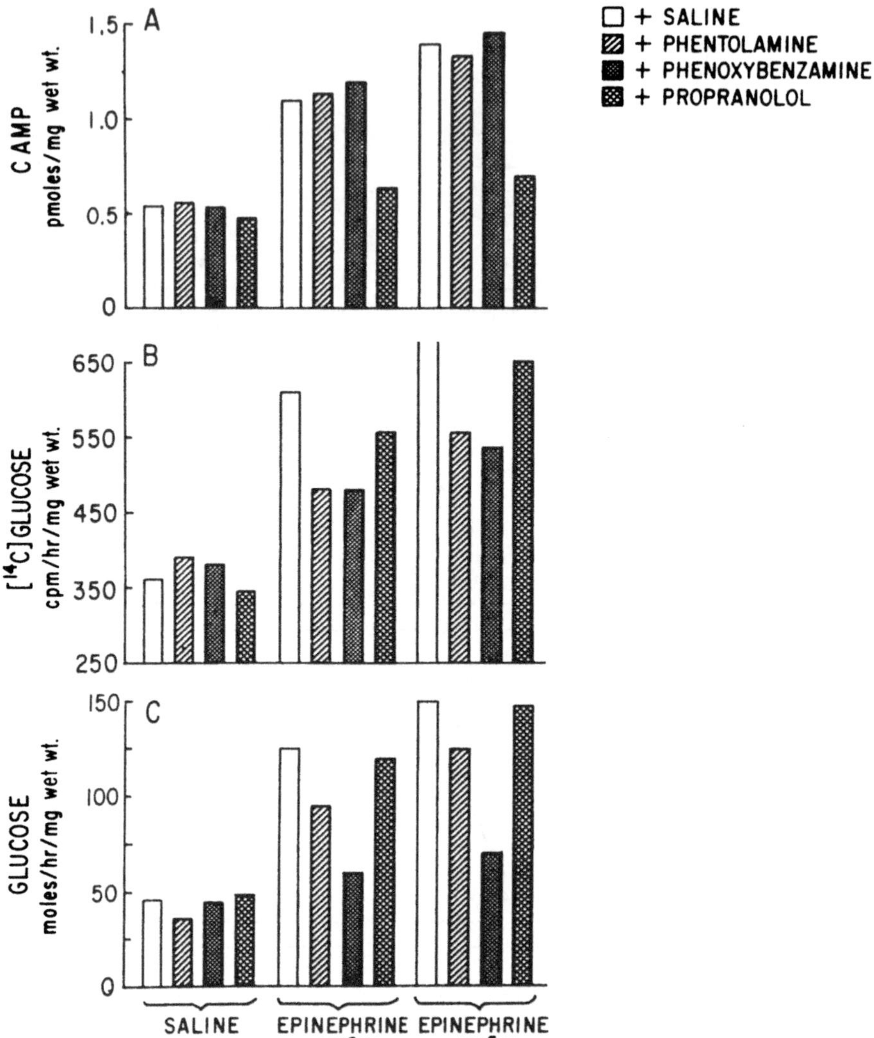


*Figure 6. Effects of α- and β-adrenergic antagonists on cyclic AMP levels (panel A), gluconeogenesis from [$^{14}C$]lactate (panel B), and glycogenolysis (panel C) in hepatocytes incubated with $10^{-6}$ M and $10^{-5}$ M epinephrine. Adrenergic blockers were presented at $10^{-5}$ M concentration. Cyclic AMP levels were measured at 1 min and glucose and [$^{14}C$] glucose at the end of 30 min. Phentolamine and phenoxybenzamine did not modify the effect of epinephrine on cyclic AMP levels at other times, whereas propranolol blocked it at all times. Data are from unpublished studies by J.H. Exton and S.C. Harper.*

Epinephrine and the specific α-adrenergic agonist phenylephrine produce changes in liver intermediary metabolites similar to those seen with glucagon (Assimacopoulos, 1976; Exton and Park, 1968), suggesting that they interact at the same sites to activate gluconeogenesis. To date, the mechanisms by which interaction of catecholamines with α-adrenergic receptors leads to activation of gluconeogenesis or glycogenolysis have not been elucidated. Pointer, Butcher, and Fain (1976) have presented evidence against possible roles for $Ca^{++}$ ions or cyclic GMP.

Epinephrine also inhibits pyruvate kinase activity in liver cells (Felui *et al.*, 1976), but it is not known whether this effect is mediated by the β-receptor-cyclic AMP system or by α-receptors. It may well result from activation of β-receptors since Chan (1976) and Jakob and Diem (1975) have found that this causes inhibition of lactate production in perfused livers or hepatocytes from fed rats whereas activation of α-receptors causes lactate release. If the inhibitory effect of epinephrine on pyruvate kinase is, in fact, mediated by β-receptors, other explanations will have to be sought for the stimulation of gluconeogenesis by this agent.

Kneer *et al.* (1974) have obtained indirect evidence in hepatocytes that epinephrine, like glucagon, inhibits flux through P-fructokinase and enhances flux through fructose diphosphatase. The inhibitory effect on P-fructokinase was unaffected by propranolol, but abolished by phenoxybenzamine or phentolamine indicating that it is mediated by α-receptors. In support of this, it was found that β-agent isoproterenol had no significant effects on glucose formation or fructose 6-P phosphorylation. Since 0.1 *mM* guanosine 3':5'-monophosphate (cyclic GMP) and related derivatives mimicked the actions of epinephrine, Kneer *et al.* (1974) suggested that the nucleotide might be the second messenger for α-adrenergic activity. However, the actions of epinephrine were also simulated by cyclic AMP and it is possible that cyclic GMP acted by raising the intracellular level of this nucleotide through competitive inhibition of its breakdown by phosphodiesterase (Beavo, Hardman, and Sutherland, 1970).

Because of the complex metabolic and hemodynamic changes resulting from administration of catecholamines *in vivo*, there have been few attempts to see if these agents directly affect hepatic glucose synthesis in intact animals. One of the problems arising from *in vitro* studies is that unphysiological concentrations ($10^{-8}$ *M* to $10^{-7}$ *M*) have to be employed to induce significant changes in glycogenolysis or gluconeogenesis (Exton and Park, 1968; Felui *et al.*, 1976; Hutson *et al.*, 1976; Pilkis, Claus, Johnson, and Park, 1975). Consequently, it has been proposed that the release of norepinephrine from nerve endings within the liver may be more important than the secretion of catecholamines from the adrenal medulla in the regulation of liver metabolism (Exton and Park,

1968). As will be discussed, the rapid stimulation of hepatic glycogenolysis and gluconeogenesis that occurs during exercise (Wahren, Felig, Ahlborg, and Jorfeldt, 1971) is probably due to increased sympathetic outflow since glucagon levels do not rise immediately.

## Role of Insulin

Insulin antagonizes the action of glucagon or catecholamines on gluconeogenesis in liver cells, but has little or no effect in the absence of these agents (Claus and Pilkis, 1976; Felui *et al.*, 1976; Pilkis *et al.*, 1975). Insulin also abolishes the changes in gluconeogenic intermediates induced by glucagon in livers perfused with alanine or lactate (Parilla *et al.*, 1976) suggesting that the hormones act at common sites. In addition, insulin inhibits the effects of glucagon, epinephrine, and cyclic AMP on pyruvate kinase (Blair, Cimbala, Foster, and Morgan, 1976; Felui *et al.*, 1976) and changes are well correlated with those in gluconeogenesis. The effect of insulin to reverse the inhibition of pyruvate kinase by glucagon is exerted rapidly and is seen at physiological levels of the hormone (Felui *et al.*, 1976). It is probably due largely to the lowering of cyclic AMP (Exton, Lewis, Ho, and Park, 1972b; Pilkis *et al.*, 1975) resulting from activation of the 'low $K_m$' form of cyclic AMP phosphodiesterase (Loten, Assimacopoulos-Jeannet, Exton, and Park, 1976).

Claus and Pilkis (1976) have presented evidence that the inhibitory effect of insulin on gluconeogenesis is also due to other factors. For example, insulin can inhibit the stimulation by the $\alpha$-adrenergic agonist phenylephrine which does not alter cyclic AMP levels (Cherrington, Assimacopoulos, Harper, Corbin, Park, and Exton, 1976). It can also suppress the effect of glucagon in the absence of extracellular $Ca^{++}$ even though cyclic AMP levels are apparently not lowered under these conditions (Claus and Pilkis, 1976). The nature of the cyclic AMP-independent mechanism(s) by which insulin may reduce gluconeogenesis is unknown.

Felig and Wahren (1971b) have shown that infusion of glucose into postabsorptive men decreases the splanchnic uptake of alanine, glycine, lactate, and pyruvate and have attributed these changes to the inhibitory action of insulin on hepatic gluconeogenesis. However, Chiasson, Liljenquist, Finger, and Lacy (1976) failed to observe an effect of physiological levels of insulin on the splanchnic extraction of alanine in dogs, although the hormone inhibited the conversion of [$^{14}$C]alanine to [$^{14}$C]glucose. Based on these observations, these authors concluded that insulin exerted its effect at one of the steps in the gluconeogenic pathway rather than on the alanine transport process. These results have recently been confirmed in man (Chiasson *et al.*, 1977).

Perhaps the clearest demonstration of the inhibitory effect of insulin on gluconeogenesis can be seen in the work of Jennings *et al.* (1977) who showed that when insulin deficiency was produced in the presence of basal levels of glucagon, gluconeogenesis increased dramatically. Figure 5 depicts experiments from their studies. Infusion of somatostatin to inhibit insulin and glucagon secretion, and of glucagon to maintain a normal IRG level, resulted in a pronounced increase in the rate of conversion of [$^{14}$C] alanine into [$^{14}$C]glucose (third panel). When basal amounts of insulin were also infused, however (fourth panel), the increase failed to occur. This study demonstrates the importance of both basal glucagon and insulin in the regulation of hepatic gluconeogenesis *in vivo*. Similar effects of endogenous glucagon and insulin on total hepatic glucose output have been reported by Cherrington, Chiasson, Liljenquist, Keller, and Lacy (1976). In addition, Liljenquist, Muller, Cherrington, Keller, Chiasson, Perry, Lacy, and Rabinowitz (1977) have recently demonstrated the importance of basal glucagon secretion in the maintenance of splanchnic glucose production in man.

## Role of Glucocorticoids

In liver of fed adrenalectomized animals, basal gluconeogenesis is not impaired (Exton, Harper, Tucker, Flagg, and Park, 1973), although there is a decreased response to glucagon or catecholamines (Exton *et al.*, 1972). In fasted or diabetic animals, adrenalectomy causes a marked reduction in hepatic gluconeogenesis (Exton *et al.*, 1973; Exton, Miller, Harper, and Park, 1976) such that rates decline to those seen in fed animals. In both instances the effects of adrenalectomy are reversed by cortisol. Thus, glucocorticoids are required for the responses of gluconeogenesis to fasting and insulin deficiency.

As illustrated in Figure 7, the effects of glucocorticoids on gluconeogenesis are evident after 1 *h* and are exerted directly on the liver, at least in part. Since they are blocked by cycloheximide (Exton *et al.*, 1973), they probably involve protein synthesis. Assays of tissue metabolites in perfused livers of diabetic rats implicate two sequences in the gluconeogenic pathway as sites of glucocorticoid action (Figure 8). These are between pyruvate and P-pyruvate and glucose-6-P and glucose. Similar observations have been made in fasted rats (Exton *et al.*, 1976).

Enzyme assays indicate that the two enzymes primarily responsible for the changes in gluconeogenic flux (and tissue metabolites) occurring with steroid depletion and repletion are P-enolpyruvate carboxykinase (Exton *et al.*, 1973; Foster, Ray, and Lardy, 1966; Shrago, Lardy, Nordlie, and Foster, 1963) and glucose 6-phosphatase (Ashmore and Weber, 1959; Nordlie, Arion, and Glende, 1965).[1]

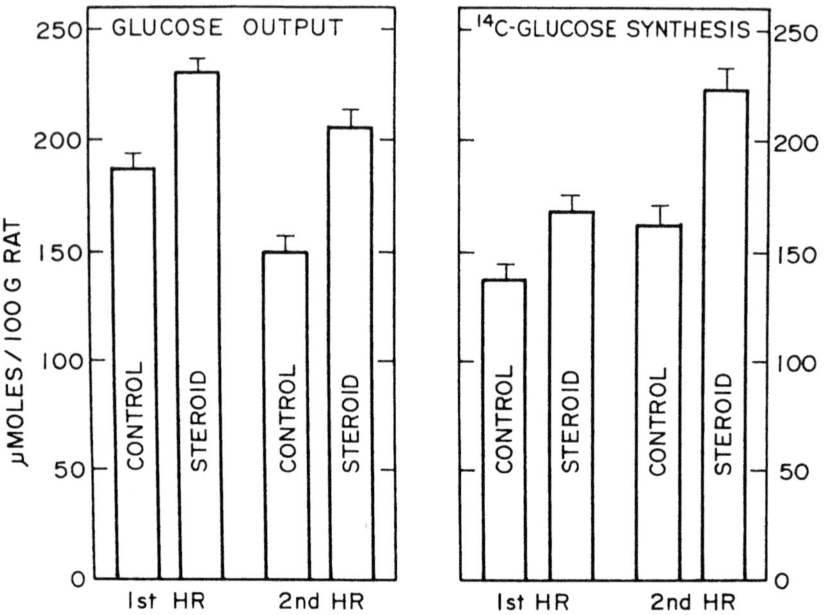


*Figure 7. Effects of cortisol in vitro on glucose production and [$^{14}C$] glucose synthesis from [$^{14}C$]lactate in perfused livers from adrenalectomized diabetic rats. Perfusions were for 2 h and the figure shows the changes during the first and second h taken separately. Data are from Exton et al. (1973).*

The induction of P-enolpyruvate carboxykinase by glucocorticoids has been studied extensively in the liver *in vivo* and in Reuber hepatoma cells. There is much evidence that it is due to increased enzyme synthesis (Barnett and Wicks, 1971; Foster *et al.*, 1966; Gunn, Hanson, Meyuhas, Reshef, and Ballard, 1975; Gunn, Tilghman, Hanson, Reshef, and Ballard, 1975) and not to decreased enzyme degradation (Gunn, Ballard, and Hanson, 1976; Gunn *et al.*, 1975). The mechanism(s) by which glucocorticoids act appears to differ from that of cyclic AMP (Barnett and Wicks, 1971) and is thought to be exerted at a transcriptional site (Wicks and McKibbin, 1972). It has also been suggested (Gunn *et al.*, 1975) that glucocorticoids act on P-enolpyruvate carboxykinase synthesis in a permissive way, *i.e.*, by increasing the sensitivity of the enzyme synthetic system to stimulation by cyclic AMP (endogenous or hormone-elevated). In support of this idea, Wicks, Barnett, and Mc-

[1]In fed rats, administration of glucocorticoids may actually decrease hepatic P-enolpyruvate carboxykinase (Ray *et al.*, 1964; Reshef, Ballard, and Hanson, 1969), but this is probably due to increased insulin secretion (Gunn *et al.*, 1975).

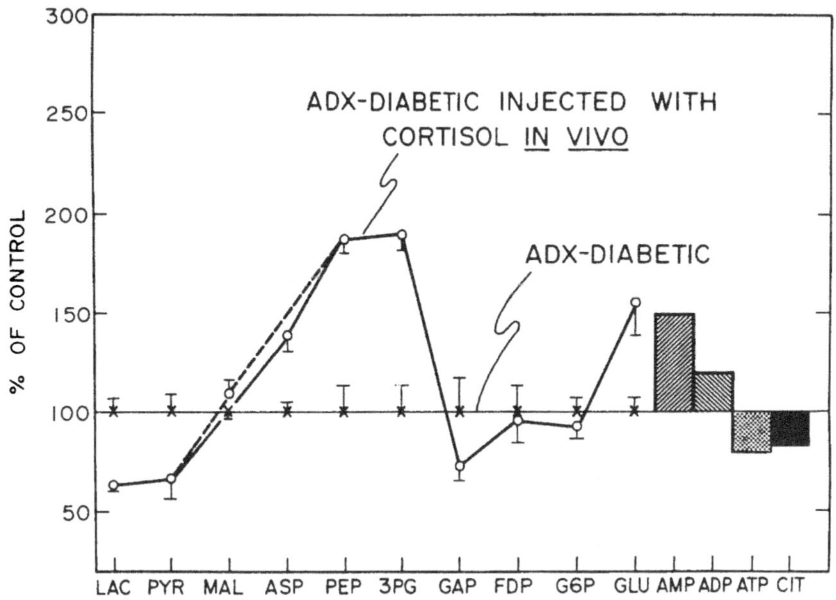


*Figure 8. Effects of cortisol treatment in vivo on the levels of gluconeogenic intermediates and adenine nucleotides in livers from adrenalectomized diabetic rats perfused for 1 h with 20 mM lactate. Cortisol sodium succinate (1 mg) was injected intraperitoneally 1 h prior to perfusion. From Exton et al (1973).*

Kibbin (1974) and Krone, Huttner, Seitz, and Tarnowski (1976) have reported that the induction of the enzyme by dibutyryl cyclic AMP is decreased in livers from adrenalectomized rats. Wicks (1971) has also observed that prior exposure of fetal explants to hydrocortisone amplifies the effects of the nucleotide. Clearly, much work is required to elucidate the interactions between glucocorticoids, insulin, and hormones which elevate cyclic AMP in the control of gluconeogenic enzyme levels.

As noted above, in addition to modulating gluconeogenesis in starvation and diabetes, glucocorticoids exert permissive effects on the actions of glucagon and catecholamines on hepatic glycogenolysis and gluconeogenesis (Exton *et al.*, 1972). Thus, livers from fasted adrenalectomized rats do not show a gluconeogenic response to these hormones (Figure 9) and neither gluconeogenesis nor glycogenolysis in livers from fed steroid-deficient rats is stimulated by low levels of these hormones. In the fasted rats, the unresponsiveness is probably due, in part, to a reduction in gluconeogenic enzymes. In the fed rats, there is an impairment in the sensitivity of gluconeogenesis and glycogenolysis to the hormones since high levels can elicit full responses. Since the ac-

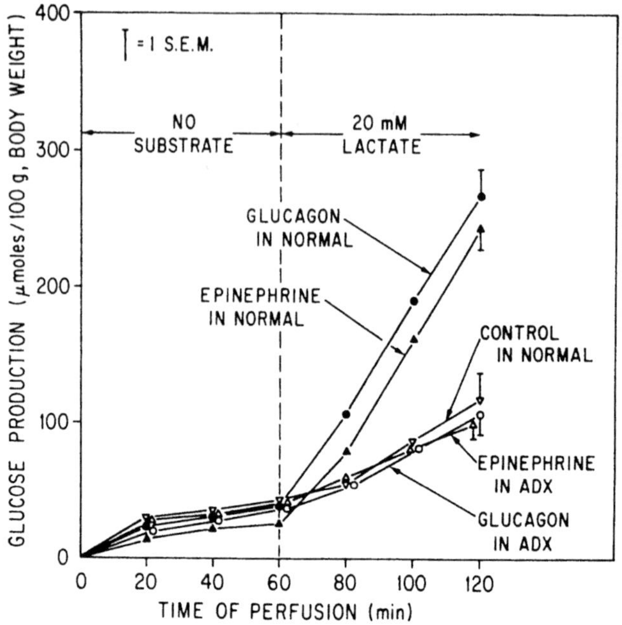


*Figure 9. Failure of glucagon or epinephrine to activate gluconeogenesis from lactate in livers from adrenalectomized rats. Livers from fasted normal or adrenalectomized rats were perfused for 1 h without substrate then lactate added to give 20* mM. *Hormones were added at 1 h to give the following concentrations: glucagon, 3 x* $10^{-8}$ M; *epinephrine,* $10^{-5}$ M. *Data from Exton et al. (1972).*

cumulation of cyclic AMP in the presence of hormones is unimpaired but the responses to exogenous cyclic AMP are diminished, it has been concluded that steroid lack impairs the sensitivity of glycogenolysis and gluconeogenesis to activation by the nucleotide (Exton *et al.*, 1972). The defect appears to be beyond the activation of protein kinase (Rousseau, Martial, and De Visscher, 1976; Zapf, Waldvogel, and Froesch, 1973). A possible role of intracellular $Ca^{++}$ and other ions in the preservation by glucocorticoids of hormonal responsiveness of enzyme systems to cyclic AMP has been suggested by Exton *et al.* (1972).

## Role of Growth Hormone

As originally shown by Russell (1957), hypophysectomy increases and growth hormone reduces the excretion of nitrogen by fasting rats. The hormone also decreases urinary urea excretion in humans (Felig, Marliss, and Cahill, 1971). In agreement with these *in vivo* findings Jefferson, Robertson, and Tolman (1973) and Tolman, Schworer, and Jefferson (1973) have found that ureogenesis, gluconeogenesis, and amino acid uptake are increased in perfused livers from hypophysectomized rats. These animals also exhibit elevated plasma urea concentrations and reduced plasma and liver amino acid levels (Tolman *et al.*, 1973). Thus, there is an enhancement of amino acid utilization for ureogenesis and gluconeogenesis in the livers of hypophysectomized rats *in vivo* and *in vitro*.

More recently, Jefferson *et al.* (1975) have reported rapid, direct effects of growth hormone to promote the uptake of cycloleucine and the incorporation of [$^{14}$C]valine into protein in perfused livers of hypophysectomized rats. In these studies the hormone also raised the intracellular concentrations of most of the glucogenic amino acids consistent with a stimulatory action on their inward transport. In apparent contradiction to earlier findings, a stimulatory effect of the hormone on ureogenesis and gluconeogenesis was also observed. However, it would seem from *in vivo* studies that this short-term effect is of less physiological importance than the long-term inhibitory action of the hormone on hepatic gluconeogenesis (Tolman *et al.*, 1973). It is not known whether the chronic inhibitory effect is exerted directly on the liver or is mediated by increased insulin secretion.

## GLUCONEOGENESIS IN DIABETES, STARVATION, AND EXERCISE

### Changes in Diabetes and Early Starvation

Diabetes in humans and experimental animals is characterized by increased gluconeogenesis with muscle protein wasting and mobilization of adipose tissue triglyceride. The mobilization of muscle amino acids in diabetes is strikingly seen in such preparations as the perfused rat hindquarter as shown in Figure 10 from the study of Ruderman and Berger (1974). The extra amino acids and glycerol released into the blood stream are converted at an accelerated rate to glucose by the liver contributing to hyperglycemia. Diabetics have decreased plasma levels of the major glucogenic acids because the increased mobilization of these amino acids from peripheral tissues is accompanied by enhanced hepatic utilization (Wahren, Felig, Cerasi, and Luft, 1972). In the case of the branched chain amino acids, hepatic uptake is minimal and these accumulate in the blood (Wahren *et al.*, 1972). In nonketotic diabetics, the kidney does not play an important role in gluconeogenesis (Wahren and Felig, 1975). However, it would be expected to contribute glucose during severe ketoacidosis.

The effects of diabetes on muscle amino acid release are attributable largely to insulin lack since they are the opposite of those exerted by the hormone. As indicated earlier, the role of

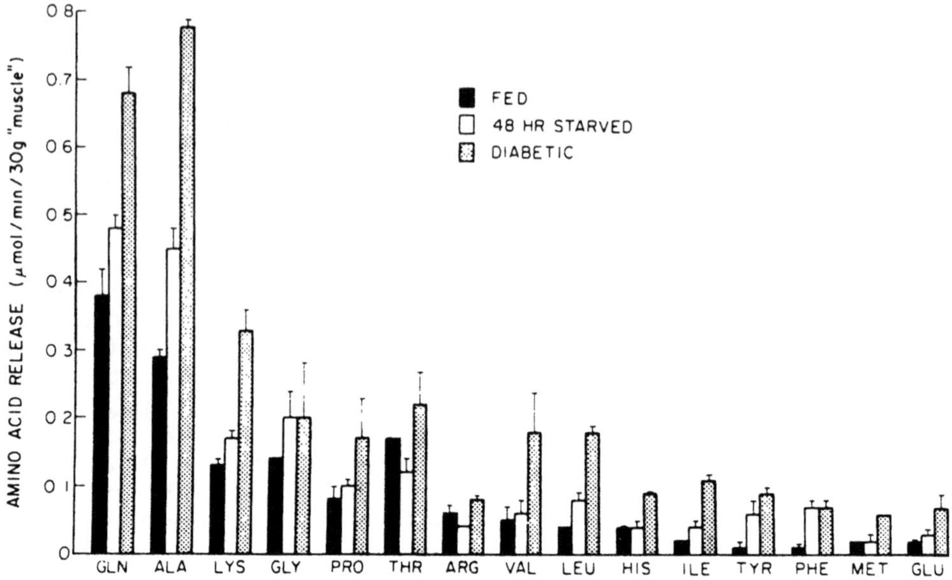


*Figure 10. Amino acid release from perfused hindquarters of fed, 48-h fasted and streptozotocin-diabetic rats. From Ruderman and Berger (1974).*

glucagon *per se* in the peripheral amino acid metabolism of normal and diabetic subjects remains uncertain.

Brief starvation in man is also associated with increased gluconeogenesis (Garber, Menzel, Boden, and Owen, 1974). Under these conditions, there is increased output of glycerol from adipose tissue and increased release of glucogenic and other amino acids from muscle, as illustrated *in vitro* in Figure 10 and *in vivo* in Figure 11 from Pozefsky, Tancredi, Moxley, Dupre, and Tobin (1976). The observed changes in peripheral metabolism are best explained by the decrease in insulin.

As alluded to earlier, studies in human diabetics have shown increased splanchnic uptake of lactate, pyruvate, alanine, and other glucogenic amino acids (see Figure 12 from Wahren *et al.*, 1972). This is due to an increased rate of hepatic gluconeogenesis which has been demonstrated many times *in vivo* and *in vitro* (for references see Exton, Corbin, and Harper, 1972). In early starvation, hepatic gluconeogenesis is also accelerated as revealed by splanchnic substrate utilization *in vivo* (Felig, Owen, Wahren, and Cahill, 1969; Garber *et al.*, 1974) and measurements of glucose synthesis in the perfused liver (Exton *et al.*. 1972). This assures that the supply of glucose to the central nervous system is maintained since ketone bodies are not a significant alternative at this stage.

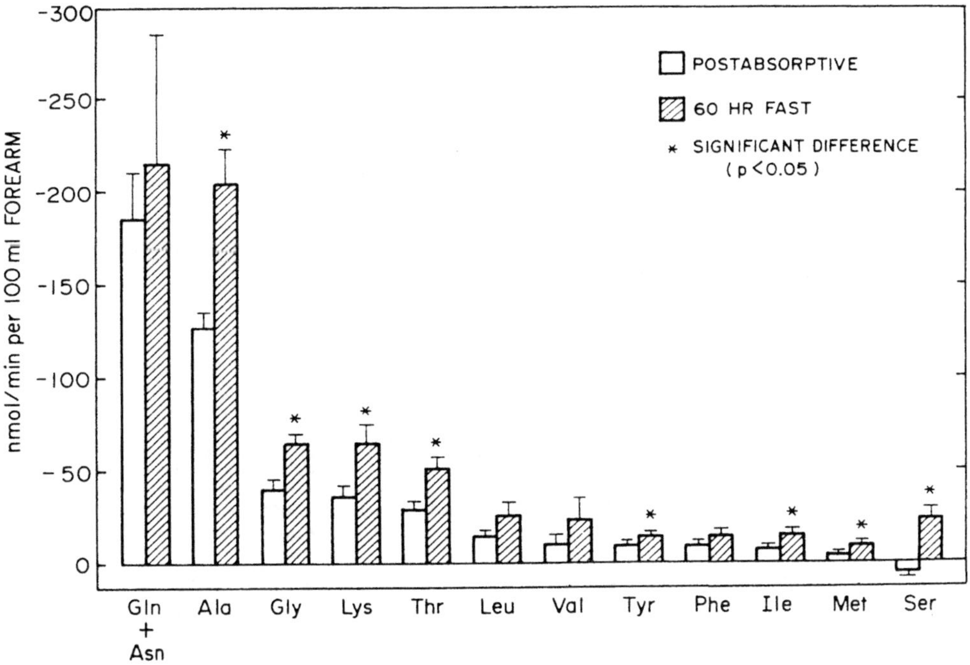


*Figure 11.* *Effect of a 60-h fast on the amino acid balance across the deep forearm tissues of man. Figure constructed from the data of Pozefsky et al. (1976).*

There is no doubt that insulin deficiency plays a major role in the increases in hepatic gluconeogenesis occurring in diabetes and early starvation. However, the role of glucagon excess remains controversial. Based on the findings of Jennings *et al.* (1977), it appears that the presence of glucagon in the insulin-deficient animal causes a rapid activation of gluconeogenesis. However, preliminary experiments by Cherrington, Liljenquist, and Chiasson (1976) have shown that deficiency of both hormones also results in a marked, but delayed, enhancement of gluconeogenesis which is inhibited by insulin. Thus, it seems that insulin deficiency alone can activate gluconeogenesis, but that glucagon causes the effect to occur sooner.

There is no doubt that alterations in enzyme levels underlie the changes in hepatic gluconeogenesis during diabetes and starvation. Many gluconeogenic enzymes are increased in activity and, in addition, glucokinase and pyruvate kinase are decreased. Measurements of tissue metabolic intermediates (Figure 13) have identified the sequences between pyruvate and P-pyruvate and between glucose-6-P and glucose as the major sites at which changes occur (for references and data see Exton, 1972). Thus, the known increases in P-enolpyruvate carboxykinase and glucose 6-phosphatase

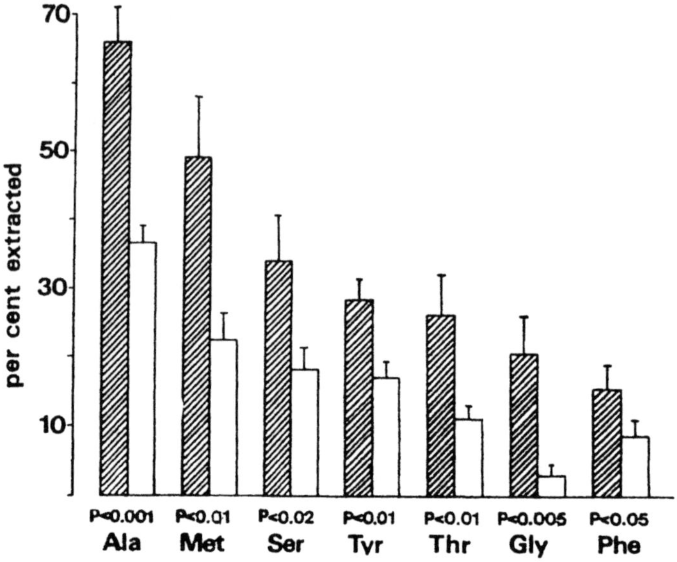


*Figure 12. Fractional extraction of amino acids by the splanchnic bed in diabetic (cross-hatched bars) and normal (open bars) subjects. From Wahren et al. (1972)*

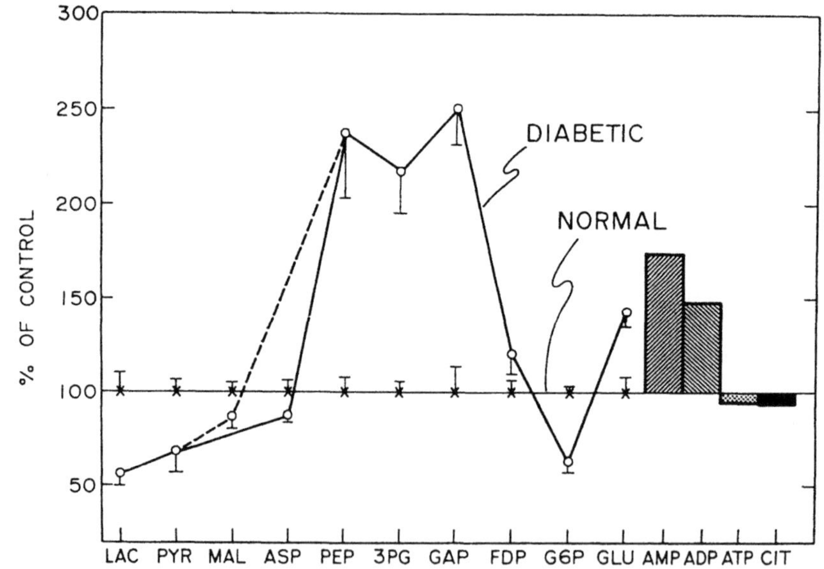


*Figure 13. Effects of alloxan diabetes on the levels of gluconeogenic intermediates and adenine nucleotides in livers perfused for 1 h with 20 mM lactate. From Exton, Harper, Tucker, and Ho (1973).*

and decreases in glucokinase or pyruvate kinase probably play significant roles. However, since metabolite analyses may not identify all the enzyme changes contributing to alterations in metabolic flux, there may be important effects at other sites.

The gluconeogenic enzyme whose synthesis and degradation has been studied most extensively in liver and which probably plays a major role in the changes in gluconeogenesis in diabetes and starvation is P-enolpyruvate carboxykinase (Hanson and Garber, 1972; Hanson, Garber, Reshef, and Ballard, 1973). The distribution of this enzyme between cytosol and mitochondria in liver differs greatly in various species. In man, guinea pig, cow, sheep, and cat, it is almost equally distributed between cytosol and mitochondria. In rat, mouse, and hamster it is mostly in the cytosol and in rabbit and pigeon mostly in the mitochondria. An important fact is that adaptive changes affect only the cytosolic form of the enzyme, which is chemically and immunologically different from the mitochondrial form. As discussed in detail by Hanson and Garber (1972), the generation of P-enolpyruvate by mitochondrial P-enolpyruvate carboxykinase may, nevertheless, play an important role in the regulation of gluconeogenesis. For example, P-enolpyruvate production in guinea pig liver mitochondria is markedly affected by changes in the $NAD^+/NADH$ ratio.

Lardy and his associates (Shrago *et al.*, 1963; Young, Shrago, and Lardy, 1964) first demonstrated the large increases in cytosolic P-enolpyruvate carboxykinase that occur during starvation and diabetes. Since inhibitors of protein synthesis block the rise in the enzyme in these conditions, it is generally assumed that increased enzyme formation is involved. A role for cyclic AMP in the induction of P-enolpyruvate carboxykinase and the enhancement of gluconeogenesis during diabetes and starvation has been proposed by Exton, Harper, Tucker, and Ho (1973). These workers found a correlation between the ability of insulin to lower cyclic AMP levels and reduce lactate gluconeogenesis and P-enolpyruvate carboxykinase activity in diabetic livers (Table 1). Krone *et al.* (1976) have also found a good correlation between liver cyclic AMP levels and P-enolpyruvate carboxykinase activity during starvation. In addition, it has been amply demonstrated that cyclic AMP or derivatives and agents which promote cyclic AMP accumulation induce several gluconeogenic enzymes including P-enolpyruvate carboxykinase in the liver *in vivo*, in liver explants or in hepatoma cells (Barnett and Wicks, 1971; Shrago *et al.*, 1963; Wicks, 1969, 1971; Wicks, Lewis, and McKibbin, 1972; Yeung and Oliver, 1968).

The induction of P-enolpyruvate carboxykinase by cyclic AMP has been shown to be due to increased enzyme synthesis (Hanson *et al.*, 1973; Wicks *et al.*, 1972). The view that the nucleotide acts mainly at a post-transcriptional site is supported by studies using

*Table 1.* *EFFECTS OF INSULIN TREATMENT ON CYCLIC AMP LEVELS, GLUCONEOGENESIS FROM [$^{14}C$]LACTATE AND P-ENOLPYRUVATE CARBOXYKINASE ACTIVITY IN LIVERS FROM DIABETIC RATS*

| Rat | Duration of insulin treatment (*hr*) | Cyclic AMP (*nmoles/g*) | [$^{14}$C]glucose synthesis (*cpm* x $10^{-3}$/*g*) | P-enolpyruvate carboxykinase (units/*g*) |
|---|---|---|---|---|
| Normal | 0.0 | 0.55 ± 0.03 | 345 ± 20 | 1.4 ± 0.1 |
| Diabetic | 0.0 | 0.96 ± 0.07 | 1130 ± 60 | 6.8 ± 0.2 |
| Diabetic | 0.5 | 0.76 ± 0.04 | 830 ± 90 | 6.1 ± 0.3 |
| Diabetic | 1.0 | 0.75 ± 0.05 | 870 ± 85 | 6.0 ± 0.3 |
| Diabetic | 2.0 | 0.68 ± 0.03 | 780 ± 90 | 5.6 ± 0.2 |
| Diabetic | 3.0 | 0.56 ± 0.02 | 500 ± 40 | 5.2 ± 0.3 |
| Diabetic | 4.0 | 0.55 ± 0.02 | 440 ± 40 | 3.9 ± 0.2 |

From Exton *et al.* (1973)

inhibitors of RNA and protein synthesis (Gunn *et al.*, 1975; Wicks and McKibbin, 1972). Blat and Loeb (1971) and Gressner and Wool (1976) have reported that glucagon and cAMP cause an increase in the phosphorylation of the 40 S ribosomal subunit protein S6. Such an effect could favor the synthesis of specific proteins by altering the binding of a component of the initiation process, for example, specific mRNA. Although cyclic AMP may also exert effects on enzyme synthesis at the level of transcription because of the phosphorylation of a specific serine residue in $f_1$ histone (Langan, 1973), there is no evidence that this plays a role in P-enolpyruvate carboxykinase synthesis.

The rapid declines in P-enolpyruvate carboxykinase seen with insulin and refeeding in diabetic and fasting rats, respectively, have been demonstrated to be mainly due to decreases in enzyme synthesis (Hopgood, Ballard, Reshef, and Hanson, 1973), although enzyme degradation also appears to increase with refeeding. Tilghman, Hanson, Reshef, Hopgood, and Ballard (1974) have shown that the fall in enzyme synthesis with refeeding and insulin can be blocked by administration of dibutyryl cyclic AMP, suggesting that it is related to a fall in liver cyclic AMP.

In analogous studies in Reuber hepatoma cells, Tilghman, Gunn, Fisher, Hanson, Reshef, and Ballard (1975) found that withdrawal of dibutyryl cyclic AMP or addition of insulin produced a rapid decline in P-enolpyruvate carboxykinase synthesis. Since enzyme synthesis could be restimulated by addition of the nucleotide in the absence of RNA synthesis, these authors concluded that deinduction was due to posttranscriptional events. Insulin also inhibited the induction of P-enolypyruvate carboxykinase by dibutyryl cyclic AMP (Barnett and Wicks, 1971; Gunn *et al.*, 1975). These findings suggest that insulin inhibits cyclic AMP action at the level of translation, perhaps by lowering the level of nucleotide, but other mechanisms are possible.

Decreased hepatic L-type pyruvate kinase activity has been observed during starvation (Krebs and Eggleston, 1965; Tanaka, Harano, Sue, and Morimura, 1967) and diabetes (Weber, Stamm, and Fisher, 1965). Although the changes probably reflect mainly alterations in total enzyme activity, they may also be due to an increase in the phosphorylation state of the enzyme since cyclic AMP is elevated in starvation and diabetes. The view that decreased pyruvate kinase activity contributes to increased gluconeogenesis is supported by the report of Friedmann *et al.* (1971) who found a marked reduction in pyruvate "recycling" in perfused livers from starved rats.

In addition to indicating a regulatory site between pyruvate and P-pyruvate, assays of tissue metabolites in fasted and diabetic livers reveal another site between glucose-6-P and glucose (Exton *et al.*, 1973; Exton and Park, 1969). This is probably due to an increase in glucose 6-phosphatase and a decrease in glucokinase. For references see Newsholme and Gevers (1967) and Steele (1966). The mechanisms by which starvation and diabetes cause an increase in glucose 6-phosphatase remain unclear. With respect to glucokinase, the decline is probably due mainly to the fall in insulin (Niemeyer, Perez, and Codoceo, 1967). A possible role for cyclic AMP in the induction of this enzyme is suggested by the finding that glucagon, catecholamines, and dibutyryl cyclic AMP prevent the increase in activity produced by glucose administration (Ureta, Radojkovic, and Niemeyer, 1970).

## Changes in Prolonged Starvation

As described in a brilliant series of investigations by Cahill and associates, hepatic gluconeogenesis declines with prolonged starvation in man (Cahill, Herrera, Morgan, Soeldner, Steinke, Levy, Reichard, and Kipnis, 1966; Felig *et al.*, 1969; Owen, Felig, Morgan, Wahren, and Cahill, 1969). This is one of the remarkable metabolic changes that the body undergoes during fasting to conserve body protein, but provide adequate fuel for vital functions.

The hormonal changes responsible for the diminution of hepatic gluconeogenesis during prolonged starvation are not immediately apparent and cannot be ascribed to the observed alterations in insulin, glucagon, or growth hormone. The present explanation is that the release of alanine and other glucogenic amino acids from muscle declines, thus, reducing the amount of substrate for hepatic gluconeogenesis (Cahill, 1970). In support of this idea Felig, Pozefsky, Marliss, and Cahill (1970) have found a marked decline in the forearm release of alanine and other glucogenic amino acids after a four- to six-week fast as illustrated in Figure 14. Consequently, the plasma concentration of alanine shows a large decrease and some other glucogenic amino acids decline (Felig *et al.*, 1969). Since studies in the perfused rat liver by Mallette *et al.*, 1969a) have shown that variations in amino acid concentrations within the physiological range greatly alter the rate of gluconeogenesis, these changes would be expected to result in a decrease in hepatic glucose synthesis. However, the large decreases in the arteriohepatic venous differences for most amino acids observed during chronic starvation, including those whose arterial concentration rises, indicates the existence of other factors whose nature remains obscure.

With respect to the decline in alanine release during prolonged fasting, Felig has provided evidence that hyperketonemia may be a contributory factor. In fasting humans, infusion of ketone bodies caused decreases in plasma alanine and urinary N excretion (Sherwin, Handler, and Felig, 1975). Addition of these compounds to incubated rat diaphragms also depressed tissue and medium levels of alanine and certain other amino acids (Palaiologos and Felig, 1976). It is

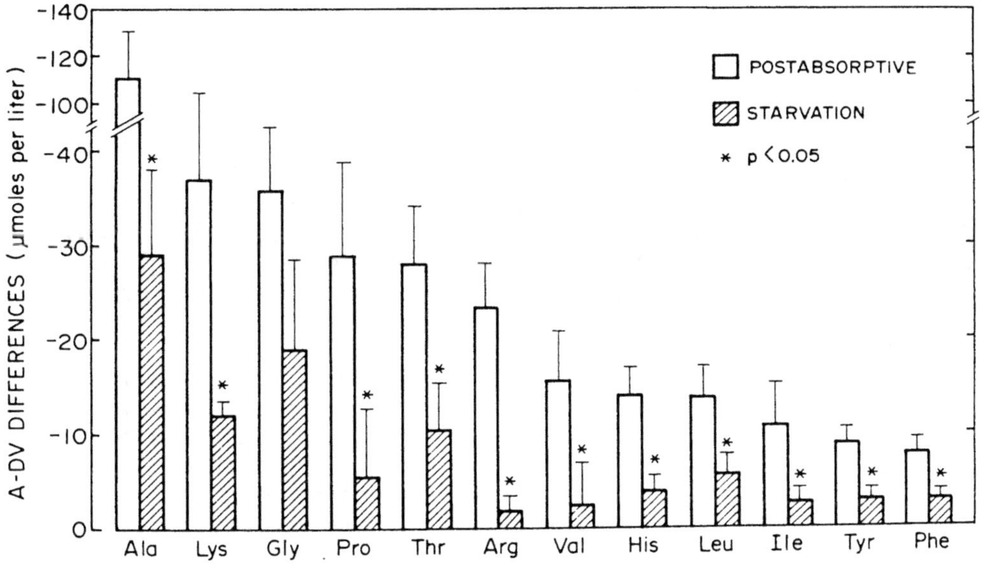


*Figure 14. Arterio-deep venous differences of plasma amino acids in the forearm of man in the postabsorptive state and after prolonged (4-6 wk) starvation. Plasma flow in the forearm was not significantly altered by starvation. Figure constructed from the data of Felig et al. (1970).*

possible that ketone bodies act by inhibiting glycolysis and/or by inhibiting branched chain amino acid catabolism.

## Changes in Exercise

Exercise results in the mobilization of lactate and alanine from muscle and glycerol and fatty acids from adipose tissue. Despite the large increase in glucose uptake occurring in muscle during exercise, the arterial concentration of this substrate remains remarkably constant. This is because hepatic glucose output is concurrently increased (Issekutz, Issekutz, and Nash, 1970; Vranic and Wrenshall, 1969), due mainly to glycogenolysis but also to gluconeogenesis, especially during prolonged exercise (Ahlborg, Felig, Hagenfeldt, Handler, and Wahren, 1974; Wahren *et al.*, 1971). Enhancement of the splanchnic uptake of lactate, pyruvate, and glycerol occurs rapidly (Wahren *et al.*, 1971) and there is also increased splanchnic fractional extraction of alanine and other glucogenic amino acids (Ahlborg *et al.*, 1974; Felig and Wahren, 1971a; Wahren, Felig, Handler, and Ahlborg, 1973). Splanchnic uptake of gluconeogenic precursors increases further during prolonged exercise and, under such conditions, gluconeogenesis can account for almost half of hepatic glucose output (Ahlborg *et al.*, 1974).

Alanine is the only amino acid consistently released from muscle during exercise and its output is related to the workload (Felig and Wahren, 1971a). With prolonged exercise alanine output increases threefold and, at this time, there is substantial uptake of leucine, isoleucine, and valine by muscle (Ahlborg *et al.*, 1974). As noted above, the catabolism of these amino acids in muscle appears to be linked to alanine production (Odessey *et al.* 1974). During recovery from exercise, the increase in leg glucose uptake and alanine release persists (Wahren *et al.*, 1973). There is also enhanced splanchnic uptake of alanine and other glucogenic amino acids due to augmented arterial levels and altered hepatic metabolism.

The increased alanine release from exercising muscle is probably largely related to the increase in glycolysis. This is shown by the linear correlation between arterial levels of alanine and pyruvate (Felig and Wahren, 1971a) and by the failure of alanine to be released by the leg muscles of exercising patients with McCardle's syndrome in which glycogen breakdown is impaired (Felig and Wahren, 1975).

The increase in muscle glycolysis during exercise appears to be mainly due to the acceleration of glycogenolysis, glucose transport, and glucose phosphorylation which occurs in working muscle. The changes in glucose transport and phosphorylation cannot be attributed to insulin since the concentration of this hormone falls with the onset of exercise and rises markedly at its cessation (Wahren *et al.*, 1973). However, the decline in insulin may contribute to the increased hepatic glucose output.

Increased sympathetic activity probably contributes to the rapid enhancement of glycogenolysis in muscle and liver during exercise, but it would not explain the increase in alanine output from muscle in view of *in vitro* findings (Li and Jefferson, 1976). The increase in glucagon during exercise (Felig *et al.*, 1972) appears to be too slow to explain the rapid rise in hepatic glycogenolysis, but it could explain the increase in gluconeogenesis occurring later and during recovery. A similar role for glucocorticoids could be postulated.

## SUMMARY AND CONCLUSIONS

1. Regulation of gluconeogenic substrate supply and modulation of the gluconeogenic pathway in the liver are both very important in the control of gluconeogenesis.

2. Insulin inhibits gluconeogenesis by reducing the release of glucogenic and other amino acids from skeletal muscle through stimulation of protein synthesis and inhibition of protein catabolism. It also inhibits the stimulatory effects of catecholamines

and glucagon on the release of glycerol from adipose tissue by lowering the intracellular level of cyclic AMP. However, its effects on muscle lactate metabolism are unclear. In addition to these peripheral effects, insulin exerts a direct effect on the liver to inhibit the stimulation of gluconeogenesis by glucagon and catecholamines. It also reverses the inhibitory effects of these hormones on pyruvate kinase. These effects are explained, at least in part, by a fall in intracellular cyclic AMP.

3. Glucagon probably does not affect muscle protein and carbohydrate metabolism directly. It is lipolytic at physiological concentrations provided insulin is absent. It rapidly stimulates gluconeogenesis and inhibits glycolysis in the liver. Both effects are mimicked by exogenous cyclic AMP. The enzymes affected lie in the sequences between pyruvate and P-pyruvate and between fructose-1, 6-di-P, and glucose-6-P. No mechanisms for cyclic AMP activation of gluconeogenic enzymes have been identified. Glucagon rapidly inactivates L-type pyruvate kinase. The effect appears to be mediated by cyclic AMP and cyclic AMP-dependent protein kinase. L-type pyruvate kinase can be phosphorylated by the latter with loss of activity and can be dephosphorylated by a histone phosphatase with recovery of activity. Changes in pyruvate kinase activity may play a major role in the regulation of gluconeogenesis by glucagon.

4. Catecholamines mobilize glycerol from adipose tissue and lactate from muscle, but inhibit amino acid release. These effects are mediated by β-adrenergic receptors. Catecholamines also rapidly activate hepatic gluconeogenesis and inhibit hepatic glycolysis. The gluconeogenic effect is exerted at the same sites as for glucagon but is mainly mediated by α-adrenergic receptors. Activation of these receptors does not result in a rise in cyclic AMP or activation of cyclic AMP-dependent protein kinase. Epinephrine rapidly inactivates L-type pyruvate kinase in liver, but it is not known whether this effect is mediated by α- or β-receptors or is related to the increase in gluconeogenesis. In contrast to the situation with glucagon and insulin, effects of catecholamines on the liver are not seen at physiological blood levels, leading to the view that the sympathetic nervous system exerts its effects through catecholamine release from nerve endings within the liver rather than from the adrenal medulla.

5. Glucocorticoids exert permissive effects on the lipolytic action of catecholamines and other hormones in adipose tissue, on the glycogenolytic action of catecholamines or glucagon in liver and skeletal muscle, and on the gluconeogenic action of catecholamines and glucagon in liver. They are not required for the cyclic AMP responses of these tissues to these hormones, but act to maintain the sensitivity of enzyme systems to activation by cyclic AMP. It is suggested that they function by preserving a normal intra-

cellular ionic environment.

Adrenal deficiency decreases the release of glucogenic and other amino acids from skeletal muscle during starvation. The effect is reversed by glucocorticoid replacement. The changes in amino acid release are accompanied by similar alterations in tissue amino acid levels and are not explained by alterations in net protein breakdown. Glucocorticoids do not alter protein catabolism and cause a small inhibition of protein synthesis. The biochemical alterations underlying the changes in amino acid metabolism induced by these steroids remain to be elucidated.

6. Growth hormone promotes amino acid transport and protein synthesis in skeletal muscle and, thus, would be expected to reduce the flow of glucogenic amino acids to the liver. A long-term inhibitory effect of growth hormone on hepatic amino acid metabolism is indicated by the increased amino acid transport, ureogenesis, and gluconeogenesis observed in livers of hypophysectomized rats and by *in vivo* changes in blood amino acid and urea levels. It is not clear whether this is exerted directly or is secondary to increased insulin secretion. Growth hormone exerts a rapid direct stimulation of amino acid transport and utilization in the liver. The physiological significance of this effect is unclear.

7. Gluconeogenesis is increased in early starvation and diabetes because of increased release of gluconeogenic substrates from peripheral tissues and because of enhancement of gluconeogenesis in the liver. Both effects are mainly attributable to insulin deficiency. Hepatic gluconeogenesis is enhanced because of increased activities of P-enolpyruvate carboxykinase and glucose 6-phosphatase and decreased activities of pyruvate kinase and glucokinase. The increase in P-enolpyruvate carboxykinase may be related to the elevation in cyclic AMP seen in starvation and diabetes since the nucleotide increases the synthesis of the enzyme apparently by acting at a posttranscriptional site. The mechanisms involved in the other enzyme changes are largely unknown. Glucocorticoids are required for the increases in gluconeogenesis in starvation and diabetes. Part of their action appears to involve modulation of P-enolpyruvate carboxykinase levels. Glucocorticoids increase the synthesis of this enzyme apparently through effects at the level of transcription.

8. Hepatic gluconeogenesis is decreased during prolonged starvation in man. This is partly due to diminished release of alanine and other glucogenic amino acids from muscle, but there is also diminished extraction of amino acids by the liver. The factors responsible for the changes in peripheral and hepatic metabolism during prolonged starvation remain unknown. It has been suggested that the decrease in muscle amino acid release is partly due to hyperketonemia.

9. Exercise causes rapid increases in muscle glucose uptake and hepatic glucose output. Muscle lactate and alanine release increases and there is enhanced splanchnic uptake of these and other gluconeogenic substrates. During prolonged exercise, alanine output from muscle increases further as does the contribution of gluconeogenesis to hepatic glucose output. In this situation, FFA account for the major part of leg muscle metabolism and branched-chain amino acids released from the splanchnic bed are taken up by muscle. During recovery from exercise, muscle glucose uptake and alanine release persist and there is augmented hepatic uptake of gluconeogenic substrates.

The changes in muscle amino acid metabolism during exercise appear to be mainly due to the effects of work. They are not attributable to insulin since its concentration declines, or to catecholamines since these agents inhibit alanine release. However, diminished insulin secretion and augmented sympathetic activity probably do contribute to the rapid increase in hepatic gluconeogenesis. The increase in glucagon which occurs later in exercise and persists through recovery may be responsible for the enhancement of gluconeogenesis seen at these times.

ACKNOWLEDGMENT

I wish to thank Dr. Alan Cherrington for his careful reading of the manuscript and very helpful comments thereon.

REFERENCES

Ahlborg, G., Felig, P., Hagenfeldt, L., Handler, R., and Wahren, J. (1974). Substrate turnover during prolonged exercise in man. *J. Clin. Invest.* 53:1080-1090.

Aoki, T.T., Muller, W.A., Brennan, M.F., and Cahill, G.F. Jr. (1974). Effect of glucagon on amino acid and nitrogen metabolism in fasting man. *Metabolism* 23:805-814.

Ashmore, J. and Weber, G. (1959). The role of hepatic glucose-6-phosphatase in the regulation of carbohydrate metabolism. *Vit. and Horm.* 17:92-132.

Assimacopoulos, F.D. (1976). Unpublished observations.

Barnett, C.A. and Wicks, W.D. (1971). Regulation of phosphoenolpyruvate carboxykinase and tyrosine transaminase in hepatoma cell cultures. I. Effects of glucocorticoids, $N^6$, $O^{2'}$-dibutyryl cyclic adenosine 3',5'-monophosphate and insulin in Reuber H35 cells. *J. Biol. Chem.* 246:7201-7206.

Beavo, J.A., Hardman, J.D., and Sutherland, E.W. (1970). Hydrolysis of guanosine and adenosine 3',5'-monophosphates by rat and bovine tissues. *J. Biol. Chem.* 245:5649-5655.

Blackshear, P.J., Holloway, P.A.H., and Alberti, K.G.M.M. (1974). The effects of starvation and insulin on the release of gluconeogenesis substrates from the extra-splanchnic tissues *in vivo*. *FEBS Lett.* 48:310-313.

Blair, J.B., Cimbala, M.A., Foster, J.L., and Morgan, R.A. (1976). Hepatic pyruvate kinase. Regulation by glucagon, cyclic adenosine 3':5'-monophosphate and insulin in the perfused rat liver. *J. Biol. Chem.* 251:3756-3762.

Blat, C. and Loeb, J.E. (1971). Effect of glucagon on phosphorylation of some rat liver ribosomal proteins *in vivo*. *FEBS Lett.* 18:124-126.

Brand, I.A. and Soling, H.D. (1975). Activation and inactivation of rat liver phosphofructokinase by phosphorylation-dephosphorylation. *FEBS Lett.* 57:163-168.

Cahill, G.F. Jr. (1970). Starvation in man. *N. Eng. J. Med.* 282: 668-675.

Cahill, G.F. Jr., Herrera, M.G., Morgan, A.P., Soeldner, J.R., Steinke, J., Levy, P.L., Reichard, G.A. Jr., and Kipnis, D.M. (1966). Hormone-fuel interrelationships during fasting. *J. Clin. Invest.* 45:1751-1769.

Caldwell, M.D., Lacy, W.W., and Exton, J.H. (1976). Unpublished observations.

Chan, T.M. (1976). Unpublished observations.

Cherrington, A.D., Assimacopoulos, F.D., Harper, S.C., Corbin, J.D., Park, C.R., and Exton, J.H. (1976). Studies on the α-adrenergic activation of hepatic glucose output. II. Investigation of the role of adenosine 3':5'-monophosphate and adenosine 3':5'-monophosphate-dependent protein kinase in the actions of phenylephrine in isolated hepatocytes. *J. Biol. Chem.* 251:5209-5218.

Cherrington, A.D., Chiasson, J.L., Liljenquist, J.E., Jennings, A.S., Keller, U., and Lacy, W.W. (1976). The role of insulin and glucagon in the regulation of basal glucose production in the post absorptive dog. *J. Clin. Invest.* 58:1407-1418.

Cherrington, A.D. and Exton, J.H. (1976). Studies on the role of cAMP-dependent protein kinase in the actions of glucagon and catecholamines on liver glycogen metabolism. *Metabolism* 25: 1351-1354.

Cherrington, A.D., Liljenquist, J.E., and Chiasson, J.L. (1976). Unpublished findings.

Chiasson, J.L., Cook, J., Liljenquist, J.E., and Lacy, W.W. (1974). Glucagon stimulation of gluconeogenesis from alanine in the intact dog. *Amer. J. Physiol.* 227:19-23.

Chiasson, J.L., Liljenquist, J.E., Cherrington, A.D., Keller, U., Sinclair-Smith, B.C., and Lacy, W.W. (1976). Unpublished findings.

Chiasson, J.L., Liljenquist, J.E., Finger, F.E., and Lacy, W.W. (1976). Differential sensitivity of glycogenolysis and gluconeogenesis to insulin infusions in dogs. *Diabetes* 25:283-291.

Chaisson, J.L., Liljenquist, J.E., Sinclair-Smith, B.C., and Lacy, W.W. (1975). Gluconeogenesis from alanine in normal postabsorptive man: Intrahepatic stimulatory effect of glucagon. *Diabetes* 24:574-584.

Clark, M.G., Kneer, N.M., Bosch, A.L., and Lardy, H.A. (1974). The fructose 1,6-diphosphatase-phosphofructokinase substrate cycle. A site of regulation of hepatic gluconeogenesis by glucagon. *J. Biol. Chem.* 249:5695-5703.

Claus, T.H. and Pilkis, S.J. (1976). Regulation by insulin of gluconeogenesis in isolated rat hepatocytes. *Biochim. Biophys. Acta* 421:246-262.

Claus, T.H., Pilkis, S.J., and Park, C.R. (1975). Stimulation by glucagon of the incorporation of U-$^{14}$C-labeled substrate into glucose by isolated hepatocytes from fed rats. *Biochim. Biophys. Acta* 404:110-123.

Ekman, P., Dahlquist, U., Humble, E., and Engstrom, L. (1976). Comparative studies on the L-type pyruvate kinase from rat liver and the enzyme phosphorylated by cyclic 3',5'-AMP-stimulated protein kinase. *Biochim. Biophys. Acta* 429:374-382.

Exton, J.H. (1972). Progress in endocrinology and metabolism Gluconeogenesis. *Metabolism* 21:945-990.

Exton, J.H., Corbin, J.G., and Harper, S.C. (1972). Control of gluconeogenesis in liver. V. Effects of fasting, diabetes and glucagon on lactate and endogenous metabolism in the perfused rat liver. *J. Biol. Chem.* 247:4996-5003.

Exton, J.H., Friedmann, N., Wong, E.H.A., Brineaux, J.P., Corbin, J.D., and Park, C.R. (1972). Interaction of glucocorticoids with glucagon and epinephrine in the control of gluconeogenesis and glycogenolysis in liver and of lipolysis in adipose tissue. *J. Biol. Chem.* 247:3579-3588.

Exton, J.H. and Harper, S.C. (1975). Role of cyclic AMP in the actions of catecholamines in hepatic carbohydrate metabolism. pp. 519-532. In G.I. Drummond, P. Greengard, and G.A. Robison (Eds.) *Advances in Cyclic Nucleotide Research*, Vol 5. Raven Press, New York.

Exton, J.H., Harper, S.C., Tucker, A.L., Flagg, J.L., and Park, C.R. (1973). Effects of adrenalectomy and glucocorticoid replacement on gluconeogenesis in perfused livers from diabetic rats. *Biochim. Biophys. Acta* 329:41-57.

Exton, J.H., Harper, S.C., Tucker, A.L., and Ho, R.J. (1973). Effects of insulin on gluconeogenesis and cyclic AMP levels in perfused livers from diabetic rats. *Biochim. Biophys. Acta* 329:23-40.

Exton, J.H., Lewis, S.B., Ho, R.J., and Park, C.R. (1972). The role of cyclic AMP in the control of hepatic glucose production by glucagon and insulin. PP. 91-101. In P. Greengard and G.A. Robison (Eds.) *Advances in Cyclic Nucleotide Research*, Vol. 1. Raven Press, New York.

Exton, J.H., Miller, T.B. Jr., Harper, S.C., and Park, C.R. (1976). Carbohydrate metabolism in perfused livers of adrenalectomized and steroid-replaced rats. *Am. J. Physiol.* 230:163-170.

Exton, J.H. and Park, C.R. (1968). Control of gluconeogenesis in liver. II. Effects of glucagon, catecholamines, and adenosine 3',5'-monophosphate on gluconeogenesis in the perfused rat liver. *J. Biol. Chem.* 243:4189-4196.

Exton, J.H. and Park, C.R. (1969). Control of gluconeogenesis in liver. III. Effects of L-lactate, pyruvate, fructose, glucagon, epinephrine, and adenosine 3',5'-monophosphate on gluconeogenic intermediates in the perfused rat liver. *J. Biol. Chem.* 244:1424-1433.

Exton, J.H., Ui, M., Lewis, S.B., and Park, C.R. (1971). Mechanism of glucagon activation of gluconeogenesis. pp. 160-178. In H-D Soling and B. Willms (Eds.) *Regulation of Gluconeogenesis*. Academic Press, New York.

Fain, J.N. and Czech, M.P. (1975). Glucocorticoid effects on lipid mobilization and adipose tissue metabolism. pp. 169-178. In H. Blaschko, G. Sayers, and A.D. Smith (Eds.) *Handbook of Physiology*, Section 7: *Endocrinology*, Vol. 6 *Adrenal gland*. American Physiological Society, Washington, DC.

Felig, P., Marliss, E.B., and Cahill, G.F. Jr. (1971). Metabolic response to human growth hormone during prolonged starvation. *J. Clin. Invest.* 50:411-421.

Felig, P., Owen, O.E., Wahren, J., and Cahill, G.F. Jr. (1969). Amino acid metabolism during prolonged starvation. *J. Clin. Invest.* 48:584-594.

Felig, P., Pozefsky, T., Marliss, E.B., and Cahill, G.F. Jr. (1970). Alanine: Key role in gluconeogenesis. *Science* 167: 1003-1004.

Felig, P. and Wahren, J. (1971a). Amino acid metabolism in exercising man. *J. Clin. Invest.* 50:2703-2714.

Felig, P. and Wahren, J. (1971b). Influence of endogenous insulin secretion on splanchnic glucose and amino acid metabolism in man. *J. Clin. Invest.* 50:1702-1711.

Felig, P. and Wahren, J. (1975). Fuel homeostasis in exercise. *N. Eng. J. Med.* 293:1078-1084.

Felig, P. Wahren, J., Hendler, R., and Ahlborg, G. (1972). Plasma glucagon levels in exercising man. *N. Eng. J. Med.* 287:184-185.

Felui, J.E., Hue, L., and Hers, H-G. (1976). Hormonal control of pyruvate kinase activity and of gluconeogenesis in isolated hepatocytes. *Proc. Nat. Acad. Sci.* 73:2762-2766.

Foster, D.O., Ray, P.D., and Lardy, H.A. (1966). Studies on the mechanisms underlying adaptive changes in rat liver phosphoenolpyruvate carboxykinase. *Biochemistry* 5:555-562.

Friedmann, B., Goodman, E.H. Jr., Saunders, H.L., Kostos, V., and Weinhouse, S. (1971). Regulation of metabolism in the liver. Estimation of pyruvate recycling during gluconeogenesis in perfused rat liver. *Metabolism* 20:2-12.

Fulks, R.M., Li, J.B., and Goldberg, A.L. (1975). Effects of insulin, glucose, and amino acids on protein turnover in rat diaphragm. *J. Biol. Chem.* 250:290-298.

Garber, A.J., Karl, I.E., and Kipnis, D.M. (1976). Alanine and glutamine synthesis and release from skeletal muscle. IV. β-adrenergic inhibition of amino acid release. *J. Biol. Chem.* 251:851-857.

Garber, A., Menzel, P.H., Boden, G., and Owen, O.E. (1974). Hepatic ketogenesis and gluconeogenesis in humans. *J. Clin. Invest.* 54:981-989.

Gerich, J.E., Lorenzi, M., Bier, D.M., Tsalikian, E., Schneider, V., Karam, J.H., and Forsham, P.H. (1976). Effects of physiological levels of glucagon and growth hormone on human carbohydrate and lipid metabolism. *J. Clin. Invest.* 57:875-884.

Gressner, A.M. and Wool, I.G. (1976). Influence of glucagon and cyclic adenosine 3':5'-monophosphate on the phosphorylation of rat liver ribosomal protein S6. *J. Biol. Chem.* 251:1500-1504.

Goodman, H.M. and Schwartz, J. (1974). Growth hormone and lipid metabolism. pp. 211-231. In E. Knobil and W.H. Sawyer (Ed.) *Handbook of Physiology*, Section 7: *Endocrinology*, Vol. 4, Part 2, *The pituitary gland and its neuroendocrine control.* American Physiological Society, Washington, DC.

Gunn, J.M., Ballard, F.J., and Hanson, R.W. (1976). Influence of hormones and medium composition on the degradation of phosphoenolpyruvate carboxykinase (GTP) and total protein in Reuber H35 cells. *J. Biol. Chem.* 251:3586-3593.

Gunn, J.M., Hanson, R.W., Meyuhas, O., Reshef, L., and Ballard, F.J. (1975). Glucocorticoids and the regulation of phosphoenolpyruvate carboxykinase (guanosine triphosphate) in the rat. *Biochem. J.* 150:195-203.

Gunn, J.M. Tilghman, S.M., Hanson, R.W., Reshef, L., and Ballard, F.J. (1975). Effects of cyclic adenosine monophosphate, dexamethasone and insulin on phosphoenolpyruvate carboxykinase synthesis in Reuber H-35 hepatoma cells. *Biochemistry* 14:2350-2357.

Hanson, R.W. and Garber, A.J. (1972). Phosphoenolpyruvate carboxykinase. I. Its role in gluconeogenesis. *Am. J. Clin. Nutr.* 25:1010-1021.

Hanson, R.W., Garber, A.J., Reshef, L., and Ballard, F.J. (1973). Phosphoenolpyruvate carboxykinase. II. Hormonal controls. *Am. J. Clin. Nutr.* 26:55-63.

Hopgood, M.F., Ballard, F.J., Reshef, L., and Hanson, R.W. (1973). Synthesis and degradation of phosphoenolpyruvate carboxylase in rat liver and adipose tissue. Changes during a starvation-refeeding cycle. *Biochemistry* 134:445-453.

Hutson, N.J., Brumley, F.T., Assimacopoulos, F.D., Harper, S.C., and Exton, J.H. (1976). Studies on the α-adrenergic activation of hepatic glucose output. I. Studies on the α-adrenergic activation of phosphorylase and gluconeogenesis and inactivation of glycogen synthase in isolated rat liver parenchymal cells. *J. Biol. Chem.* 251:5200-5208.

Issekutz, B. Jr., Issekutz, A.C., and Nash, D. (1970). Mobiliza-

tion of energy sources in exercising dogs. *J. Appl. Physiol.* 29:691-697.
Jakob, A., and Diem, S. (1975). Metabolic responses of perfused rat livers to alpha- and beta-adrenergic agonists, glucagon and cyclic AMP. *Biochim. Biophys. Acta* 404:57-66.
Jefferson, L.S., Koehler, J.O., and Morgan, H.E. (1972). Effect of insulin on protein synthesis in skeletal muscle of an isolated perfused preparation of rat hemicorpus. *Proc. Nat. Acad. Sci.* 69:816-820.
Jefferson, L.S., Li, J.B., and Rannels, S.R. (1977). Regulation by insulin of amino acid release and protein turnover in the perfused rat hemicorpus. *J. Biol. Chem.* 252:1476-1483.
Jefferson, L.S., Robertson, J.W., and Tolman, E.L. (1973). Effects of hypophysectomy on lactate metabolism in the perfused rat liver. *J. Biol. Chem.* 248:4561-4567.
Jefferson, L.S., Schworer, C.M., and Tolman, E.L. (1975). Growth hormone stimulation of amino acid transport and utilization by the perfused rat liver. *J. Biol. Chem.* 250:197-204.
Jennings, A.S., Cherrington, A.D., Liljenquist, J.E., Keller, U., Lacy, W.W., and Chiasson, J.L. (1977). The roles of insulin and glucagon in the regulation of gluconeogenesis in the postabsorptive dog. *Diabetes* 26:847-856.
Kneer, N.M., Bosch, A.L., Clark, M.G., and Lardy, H.A. (1974). Glucose inhibition of epinephrine stimulation of hepatic gluconeogenesis by blockade of the α-receptor function. *Proc. Nat. Acad. Sci.* 71:4523-4527.
Kostyo, J.L. and Nutting, D.F. (1974). Growth hormone and protein metabolism. pp. 187-210. In E. Knobil and W.W. Sawyer (Eds.) *Handbook of Physiology.* Section 7: *Endrcrinology*, Vol. 4, part 2. *The pituitary gland and its neuroendocrine control.* American Physiological Society, Washington, DC.
Krebs, H.A. and Eggleston, L.V. (1965). The role of pyruvate kinase in the regulation of gluconeogenesis. *Biochemistry* 94:30-40.
Krone, W., Hubbner, W.B., Seitz, H.J., and Tarnowski, W. (1976). Induction of rat liver phosphoenolpyruvate carboxykinase (GTP) by cyclic AMP during starvation. The permissive action of glucocorticoids. *Biochim. Biophys. Acta* 437:62-70.
Langan, T. (1973). Protein kinases and protein kinase substrates. pp. 99-153. In P. Greengard and G.A. Robison (Eds.) *Advances in Cyclic Nucleotide Research,* Vol. 3. Raven Press, New York.
Li, J.B. and Jefferson, L.S. (1977). Effect of isoproterenol on amino acid levels and protein turnover in skeletal muscle. *Am. J. Physiol.* 232:E243-249.
Liljenquist, J.E., Bomboy, J.D., Lewis, S.B., Sinclair-Smith, B.C., Felts, P.W., Lacy, W.W. Crofford, O.B., and Liddle, G.W. (1974). Effects of glucagon on lipolysis and ketogenesis in normal and diabetic men. *J. Clin. Invest.* 53:190-197.
Liljenquist, J.E., Muller, G.L., Cherrington, A.D., Keller, U., Chiasson, J.L., Perry, J.M., Lacy, W.W., and Rabinowitz, D. (1977). Evidence for an important role of glucagon in the

regulation of hepatic glucose production in normal man. *J. Clin. Invest.* **59**:369-374.
Long, C.N.H., Katzin, B., and Fry, E.G. (1940). The adrenal cortex and carbohydrate metabolism. *Endocrinology* **26**:309-344.
Loten, E.G., Assimacopoulos-Jeannet, F.D., Exton, J.H., and Park, C.R. (1976). Unpublished observations.
Ljungstrom, O., Helmquist, G., and Engstrom, L. (1974). Phosphorylation of purified rat liver pyruvate kinase by cyclic 3',5'-AMP stimulated protein kinase. *Biochim. Biophys. Acta* **358**:289-298.
Mallette, L.E., Exton, J.H., and Park, C.R. (1969a). Control of gluconeogenesis from amino acids in the perfused rat liver. *J. Biol. Chem.* **244**:5713-5723.
Mallette, L.E., Exton, J.H., and Park, C.R. (1969b). Effects of glucagon on amino acid transport and utilization in the perfused rat liver. *J. Biol. Chem.* **244**:5724-5728.
Manchester, K.L., Randle, P.J., and Young, F.G. (1959). The effect of growth hormone and of cortisol on the response of isolated rat diaphragm to the stimulating effect of insulin on glucose uptake and on incorporation of amino acids into protein. *J. Endocrinol.* **18**:395-408.
Miller, T.B. Jr., Exton, J.H., and Park, C.R. (1971). A block in epinephrine-induced glycogenolysis in hearts from adrenalectomized rats. *J. Biol. Chem.* **246**:3672-3678.
Morgan, H.E., Jefferson, L.S., Wolpert, E.B., and Rannels. D.E. (1971). Regulation of protein synthesis in heart muscle. II. Effect of amino acid levels and insulin on ribosomal aggregation. *J. Biol. Chem.* **246**:2163-2170.
Morgan, H.E., Regen, D.M., Henderson, M.J., Sawyer, T.K., and Park, C.R. (1961). Regulation of glucose uptake in muscle. VI. Effects of hypophysectomy, adrenalectomy, growth hormone, hydrocortisone, and insulin on glucose transport and phosphorylation in the perfused rat heart. *J. Biol. Chem.* **236**:2162-2168.
Narahara, H.T. and Holloszy, J.O. (1974). The actions of insulin, trypsin and electrical stimulation on amino acid transport in muscle. *J. Biol. Chem.* 249:5435-5443.
Newsholme, E.A. and Gevers, W. (1967). Control of glycolysis and gluconeogenesis in liver and kidney cortex. *Vit. and Horm.* **25**:1-87.
Niemeyer, H., Perez, N., and Codoceo, R. (1967). Liver glucokinase induction in acute and chronic insulin insufficiency in rats. *J. Biol. Chem.* **242**:860-864.
Nordlie, R.C., Arion, W.J., and Glende, E.A. Jr. (1965). Liver microsomal glucose-6-phosphatase, inorganic pyrophosphatase, and pyrophosphate-glucose phosphotransferase. IV. Effects of adrenalectomy and cortisone administration on activities assayed in the absence and presence of deoxycholate. *J. Biol. Chem.* 240:3479-3484.
Odessey, R., Khairallah, E.A., and Goldberg, A.L. (1974). Origin and possible significance of alanine production by skeletal

muscle. *J. Biol. Chem.* 249:7623-7629.

Olefsky, J.M. (1975). Effect of dexamethasone on insulin binding, glucose transport, and glucose oxidation of isolated rat adipocytes. *J. Clin. Invest.* 56:1499-1508.

Owen, O.E., Felig, P., Morgan, A.P., Wahren, J., and Cahill, G.F. Jr. (1969). Liver and kidney metabolism during prolonged starvation. *J. Clin. Invest.* 48:574-583.

Palaiologos, G. and Felig, P. (1976). Effects of ketone bodies on amino acid metabolism in isolated rat diaphragm. *Biochem. J.* 154:709-716.

Parrilla, R., Jimenez, I., and Ayuso-Parrilla, M.S. (1976). Cellular redistribution of metabolites during glucagon and insulin control of gluconeogenesis in the isolated perfused rat liver. *Arch. Biochem. Biophys.* 174:1-12.

Pilkis, S.J., Claus, T.H., Johnson, R.A., and Park, C.R. (1975). Hormonal control of cyclic 3':5'-AMP levels and gluconeogenesis in isolated hepatocytes from fed rats. *J. Biol. Chem.* 250:6328-6336.

Pilkis, S.J., Riou, J.P., and Claus, T.H. (1976). Hormonal control of [$^{14}$C] glucose synthesis from [U-$^{14}$C] dihydroxyacetone and glycerol in isolated rat hepatocytes. *J. Biol. Chem.* 251:7841-7852.

Pointer, R.H., Butcher, F.R., and Fain, J.N. (1976). Studies on the role of cyclic guanosine 3':5'-monophosphate and extracellular $Ca^{++}$ in the regulation of glycogenolysis in rat liver cells. *J. Biol. Chem.* 251:2987-2992.

Pozefsky, T., Felig, P., Tobin, J.D., Soeldner, J.S., and Cahill, G.F. Jr. (1969). Amino acid balance across tissues of the forearm in postabsorptive man. Effects of insulin at two dose levels. *J. Clin. Invest.* 48:2273-2282.

Pozefsky, T., Tancredi, R.G., Moxley, R.T., Dupre, J., and Tobin, J.D. (1976). Effects of brief starvation on muscle amino acid metabolism in nonobese man. *J. Clin. Invest.* 57:444-449.

Rannels, S.R., Li, J.B., and Jefferson, L.S. (1976). Amino acid release and protein turnover in perfused skeletal muscle of fasted normal and adrenalectomized rats. *Diabetes* 25:333.

Ray, P.D., Foster, D.O., and Lardy, H.A. (1964). Mode of action of glucocorticoids 1. Stimulation of gluconeogenesis independent of synthesis *de novo* of enzymes. *J. Biol. Chem.* 239:3396-3400.

Reshef, L., Ballard, F.J.,and Hanson, R.W. (1969). The role of the adrenals in the regulation of phosphoenolpyruvate carboxykinase of rat adipose tissue. *J. Biol. Chem.* 244:5577-5581.

Ross, B.D., Hems, R., and Krebs, H.A. (1967). The rate of gluconeogenesis from various precursors in the perfused rat liver. *Biochem. J.* 102:942-951.

Rousseau, G.G., Martial, J., and DeVisscher, M. (1976). Activity and subcellular distribution of protein kinase dependent on adenosine 3',5'-monophosphate in livers from normal and adrenalectomized rats. *Eur. J. Biochem.* 66:449-506.

Ruderman, N.B. and Berger, M. (1974). The formation of glutamine and alanine in skeletal muscle. *J. Biol. Chem.* 249:5500-5506.
Ruderman, N.B., Houghton, C.R.S., and Hems, R. (1971). Evaluation of the isolated perfused rat hindquarter for the study of muscle metabolism. *Biochem. J.* 124:639-651.
Russell, J.A. (1957). Effects of growth hormone on protein and carbohydrate metabolism. *Am. J. Clin. Nutr.* 5:404-416.
Sanders, R.B. and Riggs, T.R. (1967). Effects of epinephrine on the distribution of two model amino acids in the rat. *Mol. Pharmacol.* 3:352-358.
Scrutton, M.C. and Utter, M.F. (1968). The regulation of glycolysis and gluconeogenesis in animal tissues. *Ann. Rev. Biochem.* 37:249-302.
Sherwin, R.S., Handler, R.G., and Felig, P. (1975). Effect of ketone infusions on amino acid and nitrogen metabolism in man. *J. Clin. Invest.* 55:1382-1390.
Shrago, E., Lardy, H.A., Nordlie, R.C., and Foster, D.O. (1963). Metabolic and hormonal control of phosphoenolpyruvate carboxykinase and malic enzyme in rat liver. *J. Biol. Chem.* 238:3188-3192.
Smith, O.K. and Long, C.N.H. (1967). Effect of cortisol on the plasma amino nitrogen of eviscerated adrenalectomized-diabetic rats. *Endocrinology* 80:561-566.
Soderling, T.R., Corbin, J.D., and Park, C.R. (1973). Regulation of adenosine 3':5'-monophosphate-dependent protein kinase. II. Hormonal regulation of the adipose tissue enzyme. *J. Biol. Chem.* 248:1822-1829.
Soderling, T.R. and Park, C.R. (1974). Recent advances in glycogen metabolism. pp. 283-333. In P. Greengard and G.A. Robison (Eds.) *Advances in Cyclic Nucleotide Research*, Vol. 4. Raven Press, New York.
Steele, R. (1975). Influences of corticosteroids on protein and carbohydrate metabolism. pp. 136-167. In H. Blaschko, G. Sayers, and A.D. Smith (Eds.) *Handbook of Physiology.* Section 7: *Endocrinology*, Vol. 6. *Adrenal Gland.* American Physiological Society, Washington, DC.
Steele, R. (1966). The influences of insulin on the hepatic metabolism of glucose. *Ergeb. Physiol.* 57:91-189.
Tanaka, T., Harana, Y., Sue, F., and Morimura, H. (1967). Crystallization, characterization and metabolic regulation of two types of pyruvate kinase isolated from rat tissues. *J. Biochem.* (Tokyo) 62:71-91.
Theinhaus, R., Tharandt, L., Zais, U., and Staib, W. (1975). Einfluss von glucocorticoiden auf die freisetzung von aminosauren des perfundierten hinterkorpers adrenalecktomierter ratten. *Hoppe-Seyler's Z. Physiol. Chem.* 356:811-817.
Tilghman, S.M., Gunn, J.M., Fisher, L.M., Hanson, R.W., Reshef, L., and Ballard, F.J. (1975). Deinduction of phosphoenolpyruvate carboxykinase (guanosine triphosphate) synthesis in Reuber H-35 cells. *J. Biol. Chem.* 250:3322-3329.

Tilghman, S., Hanson, R.W., Reshef, L, Hopgood, M.F., and Ballard, F.J. (1974). Rapid loss of translatable messenger RNA of phosphoenolpyruvate carboxykinase during glucose repression in liver. *Proc. Nat. Acad. Sci.* 71:1304-1308.

Titanji, V.P.K., Zetterquist, P., and Engstrom, L. (1976). Regulation *in vitro* of rat liver pyruvate kinase by phosphorylation-dephosphorylation reactions, catalyzed by cyclic AMP dependent protein kinases and a histone phosphatase. *Biochim. Biophys. Acta* 422:98-108.

Tolbert, M.E.M., Butcher, F.R., and Fain, J.H. (1973). Lack of correlation between catecholamine effects on cyclic adenosine 3':5'-monophosphate and gluconeogenesis in isolated rat liver cells. *J. Biol. Chem.* 248:5686-5692.

Tolman, E.L., Schworer, C.M., and Jefferson, L.S. (1973). Effects of hypophysectomy on amino acid metabolism and gluconeogenesis in the perfused rat liver. *J. Biol. Chem.* 248:4552-4560.

Ui, M., Claus, T.H., Exton, J.H., and Park, C.R. (1973). Studies on the mechanism of action of glucagon on gluconeogenesis. *J. Biol. Chem.* 248:5344-5349.

Ui, M., Exton, J.H., and Park, C.R. (1973). Effects of glucagon on glutamate metabolism in the perfused rat liver. *J. Biol. Chem.* 248:5350-5359.

Ureta, T., Radojkovic, J., and Niemeyer, H. (1970). Inhibition by catecholamines of the induction of rat liver glucokinase. *J. Biol. Chem.* 245:4819-4824.

Veneziale, C.M. (1971). Gluconeogenesis from fructose in isolated rat liver. Stimulation by glucagon. *Biochemistry* 10:3443-3447.

Vranic, M. and Wrenshall, G.A. (1969). Exercise, insulin, and glucose turnover in dogs. *Endocrinology* 85:165-171.

Wahren, J. and Felig, P. (1975). Renal substrate exchange in human diabetes mellitus. *Diabetes* 24:730-734.

Wahren, J., Felig, P., Ahlborg, G., and Jorfeldt, L. (1971). Glucose metabolism during leg exercise in man. *J. Clin. Invest.* 50:2715-2725.

Wahren, J., Felig, P., Cerasi, E., and Luft, R. (1972). Splanchnic and peripheral glucose and amino acid metabolism in diabetes mellitus. *J. Clin. Invest.* 51:1870-1878.

Wahren, J., Felig, P., Hendler, R., and Ahlborg, G. (1973). Glucose and amino acid metabolism during recovery after exercise. *J. Appl. Physiol.* 34:838-845.

Weber, G., Stamm, N.B., and Fisher, E.A. (1965). Insulin: Inducer of pyruvate kinase. *Science* 149:65-67.

Wicks, W.D. (1971). Differential effects of glucocorticoids and adenosine 3':5'-monophosphate on hepatic enzyme synthesis. *J. Biol. Chem.* 246:217-233.

Wicks, W.D. (1969). Induction of hepatic enzymes by adenosine 3',5'-monophosphate in organ culture. *J. Biol. Chem.* 244:3941-3950.

Wicks, W.D., Lewis, W., and McKibbin, J.B. (1974). Interaction between hormones and cyclic AMP in regulating specific hepatic enzyme synthesis. *Fed. Proc.* 33:1105-1111.

Wicks, W.D., Lewis, W., and McKibbin, J.B. (1972). Induction of phosphoenolpyruvate carboxykinase by $N^6,O^{2'}$-dibutyryl cyclic AMP in rat liver. *Biochim. Biophys. Acta* 264:177-185.

Wicks, W.D. and McKibbin, J.B. (1972). Evidence for translational regulation of specific enzyme synthesis by $N^6,O^{2'}$-dibutyryl cyclic AMP in hepatoma cell cultures. *Biochem. Biophys. Res. Commun.* 48:205-211.

Williamson, J.R. (1975). Effects of epinephrine on glycogenolysis and myocardial contractility. pp. 605-636. In H. Blaschko, G. Sayers, and A.D. Smith (Eds.) *Handbook of Physiology.* Section 7: *Endocrinology*, Vol. 6, *Adrenal Gland.* American Physiological Society, Washington, DC.

Yeung, D. and Oliver, I.T. (1968). Induction of phosphopyruvate carboxylase in neonatal rat liver by adenosine 3',5'-cyclic monophosphate. *Biochemistry* 7:3231-3239.

Young, J.W., Shrago, E., and Lardy, H.A. (1964). Metabolic control of enzymes involved in lipogenesis and gluconeogenesis. *Biochemistry* 3:1687-1692.

Zapf, J., Waldvogel, M., and Froesch, E.R. (1973). Protein kinase and cyclic AMP-binding activities in liver and adipose tissue of normal streptozotocin-diabetic and adrenalectomized rats. *FEBS Lett.* 36:253-256.

# ENERGY METABOLISM IN FEASTING AND FASTING

O.E. Owen, G.A. Reichard, Jr., M.S. Patel, and G. Boden

Temple University Health Sciences Center
3401 N. Broad Street
Philadelphia, PA 19140

## OUTLINE

I. Introduction

II. Feasting and Fasting

III. Interrelationship among glycogenolysis, gluconeogenesis, and ketogenesis during fasting

IV. Relationship between free fatty acid concentration and utilization

V. Relationship between ketone-body concentrations and utilization

VI. Ketone-body oxidation and acid-base balance

VII. Summary

## INTRODUCTION

The history of humans is, in large part, a chronicle of their quest for food. Humans can feast when food is available and fast during caloric deprivation without serious consequences. The capability of humans to survive marked extremes in caloric intake depends on the ability (1) to selectively oxidize and excrete certain nitrogenous metabolites, (2) to store carbohydrate and lipid metabolites in an economical form during caloric excess, and (3) to spare protein and to selectively mobilize and oxidize substrates derived from fat stores during caloric deprivation.

*Table 1. PHYSIOLOGIC RESPONSES TO FEASTING AND PROLONGED FASTING*

| | Feasting | Prolonged fast |
|---|---|---|
| Resting metabolic rate | 110% | 85% |
| Body temperature | 99°F | 96-97°F |
| Respiratory quotient | 1.00 | 0.68 |

## FEASTING AND FASTING

There is a tendency to compare alterations in metabolism in the postabsorptive state (overnight fast) with those during the prolonged starvation state. However, changes in different nutritional states become more obvious when the feasting state is contrasted with the prolonged fasting state. Table 1 depicts the physiologic responses to these two contrasting states.

During feasting on a carbohydrate, fat, and protein meal there is an increase in resting metabolic rate, body temperature, and respiratory quotient (Goldman, Haisman, Bynum, Horton, and Sims, 1973). These increases are most obvious when the human goes from the prolonged starving state to the feasting state. During prolonged starvation the resting metabolic rate progressively diminishes to about 85% of the predicted basal metabolic rate (Benedict, 1915), while during feasting the resting metabolic rate increases to about 110% of the predicted basal metabolic rate (Goldman *et al.*, 1973). After prolonged starvation the body temperature falls to about 96-97°F (Owen, unpublished observation) while during feasting it increases to about 99°F (Goldman *et al.*, 1973). The differences in body temperature between feasting and fasting can, at least in part, be related to differences in the metabolic rate. A rough rule is that the metabolic rate increases about 10% for each degree increase in body temperature (DuBois, 1921).

The nonprotein respiratory quotient theoretically should be 0.70 when fatty acid utilization is maximum during starvation. However, during prolonged starvation the nonprotein respiratory quotient is about 0.68 (Owen and Morgan, unpublished observation). The reason for this low respiratory quotient is that some of the oxygen consumed for *beta*-oxidation of long-chain fatty acids in the liver produces ketone bodies which are lost in the urine. About 8% of the ketone bodies produced by hepatic ketogenesis is unaccompanied by $CO_2$ production because of ketonuria. This results in a total body nonprotein respiratory quotient of less than 0.70.

Providing carbohydrate and protein are consumed as well as fat, feasting increases the respiratory quotient. If the diet is predominantly carbohydrate, the respiratory quotient approaches, or exceeds, 1.0 (Goldman *et al.*, 1973; Passmore and Swindells, 1963).

The fuels the body needs for energy requirements can be derived from exogenous (food) or endogenous (tissue stores) sources. The average human can be encouraged to consume no more than 8,000-12,000 *Kcal* per day. Caloric quantities in excess of this are not readily consumed because of marked anorexia. However, the amount of time required to ingest food either to replenish or increase body stores is of interest. A hungry adult human can readily eat 5-10 *Kcal* per pound body weight in 10 *min*. Thus, an individual weighing 150 *lbs* can eat approximately one-half to all (750-1500) his daily basal caloric requirements in 10 *min*. By contrast, the delivery of calories from stored fuels during the fasting resting state by the processes of lipolysis, glycogenolysis, gluconeogenesis, ketogenesis, and transiently from proteolysis amounts to about 1 *Kcal* per 150 *lbs* body weight per *min*. Thus, per unit of time, exogenous food can normally provide a 75- to 150-fold greater supply of calories than is needed in the resting state. This is similar to what is observed with the automobile. It takes only minutes to fill the gas tank, but it takes hours of use to consume a full tank of gas. However in humans, digestion, absorption, conversion, oxidation, synthesis, and excretion of the ingested food permit only minor perturbation in bloodstream metabolites.

Insulin is the only hormone regulating the flow of fuels that consistently increases in the bloodstream following the ingestion of food (Table 2). Plasma insulin concentration increases two- to

*Table 2. HORMONAL, SUBSTRATE, AND DEPOT RESPONSES TO FEASTING*

| | Responses |
|---|---|
| HORMONE | |
| Insulin (storage hormone) | Increases 2-6 fold |
| SUBSTRATE | |
| Glucose | Increases 1.5 fold |
| Amino acids | Increases 1.5 fold |
| Triglycerides | Increases 2 fold |
| Free fatty acids | Decreases 2 fold |
| DEPOT | |
| Protein | Trivial or replete |
| Glycogen | Replete |
| Triglycerides | Replete or expand |

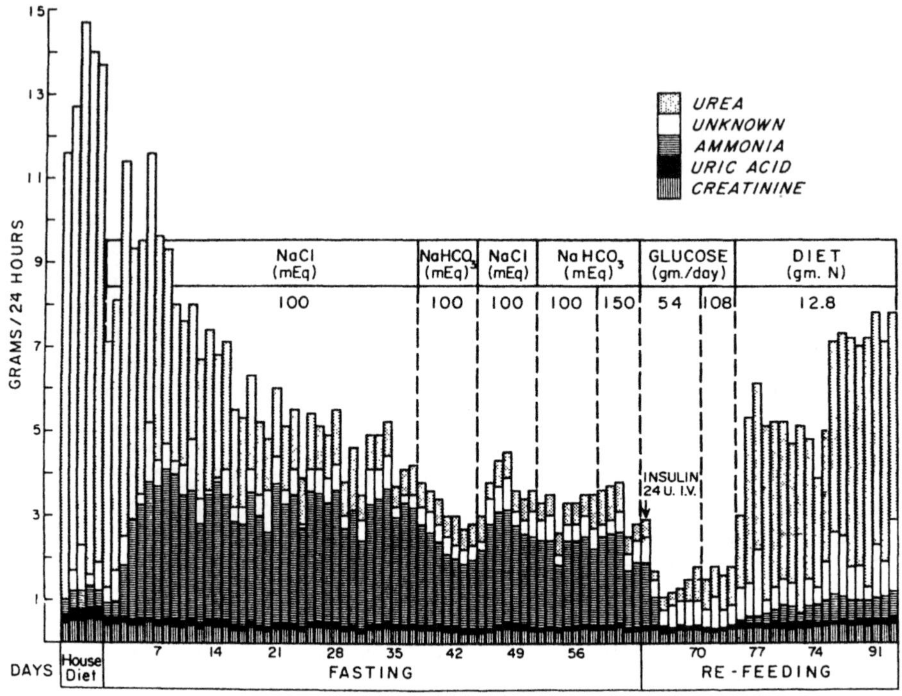


*Figure 1. Daily urinary nitrogen excretion in a female patient fasted for 64 days. During the latter part of starvation she received $NaHCO_3$ alternately with NaCl. Intravenous insulin was followed by oral glucose administration. Daily diet (days 75-93) consisted of 6.5 Kcal/lb body weight containing 12.8 g of protein nitrogen.*

six-fold following a meal balanced in carbohydrate, fat and protein (Ahmed, Gannon, and Nuttall, 1976), a meat meal (Felig, Wahren, and Hendler, 1975; Müller, Faloona, Aguilar-Parada, and Unger, 1970), or a glucose Load (Felig *et al.*, 1975; Müller *et al.*, 1970). The response of glucagon varies and depends on the type of food ingested. The circulating concentration of glucagon increases after a protein meal (Müller *et al.*, 1970), decreases after a carbohydrate meal (Müller *et al.*, 1970), and remains stable after a balanced meal (Ahmed *et al.*, 1976). A meat meal causes total amino acid concentration to increase about 1.5-fold and causes free fatty acid concentration to decrease two-fold (Wahren, Felig, and Hagenfeldt, 1976). Unfortunately, the published information on circulating hormonal and substrate responses to a balanced meal is limited.

Although the accumulation of catabolic nitrogenous metabolites is incompatible with life, conservation of limited body protein (nitrogen) mass during starvation is essential for survival. During total starvation daily nitrogen excretion decreases, transiently increases and then progressively decreases for weeks to a fairly constant rate of 4-5 *g*/day (Figure 1). Early attempts in humans to decrease nitrogen wasting during starvation by giving sodium bicarbonate (Owen and Cahill, unpublished observation) were not as effective as administering *alpha* ketoanalogues of essential amino acids (Sapir, Owen, Pozefsky, and Walser, 1974). However, Figure 1 shows that (1) during total starvation for 64 days this subject lost about 341 *g* of urinary nitrogen or about 1/3 of the entire nitrogen mass; (2) intravenous insulin and subsequent oral glucose resulted in sparing of body nitrogen; and (3) while consuming a hypocaloric diet of 6.5 *Kcal* per pound of body weight per day containing 12.8 *g* of nitrogen, about 131 *g* of nitrogen was gained in 19 days. Thus, while consuming a diet adequate in protein but inadequate in calories to maintaining adipose tissue stores, a positive nitrogen balance occurred to replenish lean body tissues.

There have been published reports stressing the importance of branched-chain amino acids as fuels for muscle metabolism (Adibi, 1976) and alanine (Felig, 1973) and glutamine (Marliss, Aoki, Pozefsky, Most, and Cahill, 1971) as substrates for gluconeogenesis. Quantitatively, however, amino acids are minor substrates for either process during starvation. From the nitrogen excretion data shown in Figure 1, and the fact that the urea does not accumulate in the body during starvation, it is obvious that amino acids are minor energy-yielding substrates during prolonged fasting. It should be recognized, however, that there are important distinctions between physiologic and quantitative contributions of amino acids as energy-yielding substrates. This is exemplified by reviewing the role of alanine in maintaining glucose homeostasis during starvation. It is well established that alanine is the primary gluconeogenic amino acid during starvation, accounting for 41%-48% of the total splanchnic extraction of amino acids (Owen, Patel, Block, Kreulen, Reichle, and Mozzoli, 1976). However, the contribution of alanine to hepatic glucose production during prolonged starvation is only about 10 *g*/day or 10% of total glucose production (Felig, Owen, Wahren, and Cahill, 1969; Garber, Menzel, Boden, and Owen, 1974). Quantitatively this is not much glucose. Nevertheless, the contribution of alanine to glucose homeostasis can be critical during starvation. When blood glucose concentration in starving humans is 65-75 *mg/dℓ*, there are only about 11-14 *g* of glucose in the entire body of an adult man (Cahill, Herrera, Morgan, Soeldner, Steinke, Levy, Reichard, and Kipnis, 1966). If glucose utilization and the contribution of lactate, pyruvate, glycerol, and amino acids other than alanine to gluconeogenesis remained constant during prolonged starvation, while the contribution of alanine to gluconeogenesis ceased, catastrophic hypoglycemic coma would be induced in less than one day of additional starvation.

In contrast to the potential massive caloric store of fat as triglycerides in adipose tissue, and the much smaller, yet significant, stores of carbohydrate as glucose and glycogen, there is no specific depot for proteins or amino acids (nitrogen). Surplus dietary amino acids are either converted to carbohydrate or to lipid, or oxidized as fuel, and the nitrogen excreted predominantly as urinary urea. Because of the lack of a storage depot, excess amino acids not used for protein synthesis are degraded. It has been shown that when an already adequate human diet is supplemented with an additional 320 *g* of meat, 56%, 26%, and 18% of the extra nitrogen is eliminated, respectively, on the subsequent three days (Krebs, 1972). Krebs has attributed selective degradation of amino acids to the abundance of amino acid catabolizing enzymes which are responsive to nutritional alterations (Krebs, 1972).

The cardinal role of the liver in maintaining fuel homeostasis is well recognized. Felig *et al.* (1975) used direct catheterization techniques to evaluate the role of the splanchnic (hepatic) and peripheral tissues in the disposal of 100 *g* of ingested glucose. They estimated that 85 *g* (85%) of the ingested glucose was extracted by the liver. Although some of the glucose was probably converted to glycogen, $CO_2$ and $H_2O$, the metabolic fate of the glucose extracted by the liver was not quantitatively determined. However, some reasonable speculation based on published data can be made.

After an overnight fast the mean glycogen content of normal human liver is 43.7 *g/kg* wet weight tissue (Nilsson, 1973), and it can be increased to only 58.9 *g/kg* wet weight tissue following a four-hour period during which about 4 *g* of glucose per *kg* body weight (280 *g*/70 *kg*/4 *h*) is infused intravenously (Nilsson and Hultman, 1974). Although the deposition of glucose as glycogen in various tissues may be dependent on the route of administration, this load of glucose should be sufficient to induce maximum hepatic glycogen storage. If it can be assumed that 100 *g* oral glucose load produces a mean hepatic glycogen content similar to that observed after about 280 *g* of glucose given intravenously, calculations indicate that approximately 20 *g* of the orally-administered glucose is converted to and stored as hepatic glycogen.* Thus, of the 85 *g* of glucose extracted by the liver only 20 *g* are converted to glycogen and the remaining glucose must be oxidized to $CO_2$ and $H_2O$ or converted to other products.

Although data in the literature pertaining to the conversion of glucose to lipid in adipose tissue and liver are ambiguous, recent reports suggest that in humans the liver may be the site for this process. When humans consume a eucaloric diet in which fat

*Adult man liver weighs 1500 *g* (Sherlock, 1975). Therefore, 58.9 *g/kg*-43.7 *g/kg* x 1.5 *kg* = 22.8 *g*. The conversion factor for glucose to glycogen is 0.9. Thus, 22.8 *g* x 9/10 = 20.5 *g*.

provides 40%-45%, carbohydrate 45%-50%, and protein the remainder of the caloric intake, there may be little need to synthesize fat from glucose and/or proteins (Owen and Reichard, 1971). Barter, Nestel, and Carroll (1972) presented good evidence that, in healthy humans eating diets containing approximately 40% of calories as carbohydrates, 40% as fat, and 20% as protein, the newly-formed plasma triglyceride fatty acids in the postabsorptive period were predominantly derived from hepatic esterification of plasma free fatty acids. However, when excessive calories are provided, Keys, Anderson, and Brozek (1955) pointed out that the increase in body weight is almost entirely due to an increase in body fat. This increase in fat mass need not simply be due to storage of exogenously derived lipids. Wilmore and Dudrick (1968) demonstrated that excessive provision of glucose, amino acids, vitamins, and minerals by intravenous nutrition to children resulted in an anabolic state in which lean body mass and lipid stores were increased. The site for this *de novo* synthesis of lipids in the human has been controversial. Barter *et al.* (1972) showed that when normal subjects were fed diets in which glucose provided 80%-85%, fat 5%-10%, and protein 5%-10% of the calories, glucose makes a significant contribution to the fatty acid moiety of endogenously-formed triglycerides synthesized in the liver, released into the bloodstream, packaged as very low-density lipoprotein, and destined for peripheral tissue removal (Boberg, Carlson, Freyschuss, Lassers, and Wahlquist, 1972; Havel, Kane, Balasse, Segel, and Basso, 1970; Owen and Reichard, 1971). The results of Barter *et al.* (1972) together with the data of Patel, Owen, Goldman, and Hanson (1975) strongly support the contention that in humans the liver is the predominant site for *de novo* fatty acid synthesis. Thus, it appears that the liver adds glucose to the bloodstream via glycogenolysis and gluconeogenesis during fasting and extracts glucose from the bloodstream for glycogen and lipid syntheses during feasting.

The triglycerides synthesized in the liver and released as very low-density lipoproteins are relatively unimportant in terms of direct energy delivery to tissues, but are important for the transport of free fatty acids through the bloodstream for peripheral storage as triglycerides in adipose tissue.

During feasting on a mixed meal, the major portion of plasma triglycerides are derived from fat absorption from the gut. After partial hydrolysis in the small intestine, fat is absorbed and transported in the lymph as chylomicrons, which are oil droplets of triglycerides surrounded by an amphiphilic coat of phospholipids and proteins (Havel, 1972). The lymphatic route bypasses the liver, and the bulk of the chylomicron triglyceride is carried to the various capillary beds, most of which are situated in adipose tissue and muscle. The glucose that is not extracted by the liver after eating produces postprandial hyperglycemia and triggers insulin secretion. Insulin promotes the uptake of glucose by adipose tissue and thereby increases the formation of *alpha*-glycerophosphate for esterification of free fatty acids. Insulin also activates the synthesis of lipoprotein lipase, the enzyme that hydrolyses chylomicron triglycerides and permits the entry of free fatty acids into cells for storage. In addition, insulin inhibits the intracellular hormone-sensitive lipase in adipose tissue and, thus, prevents the formation and release of intracellular free fatty acids into the plasma. Thus, uptake and storage of plasma triglycerides are promoted while mobilization of intracellular triglycerides is decreased. It has been estimated that a 70 *kg* man with 14 *kg* of adipose tissue can account for an uptake of about 56 *g* of chylomicron triglycerides per hour (Owen and Reichard, 1971). This can result in a large expansion of the adipose tissue mass.

## INTERRELATIONSHIPS AMONG GLYCOGENOLYSIS, GLUCONEOGENESIS, AND KETOGENESIS DURING FASTING

The liver, kidney, adipose tissue, and muscle have an excellent working relationship during fasting; these organs supply, convert, and conserve fuels required by the rest of the body. During the postabsorptive fast the liver is transiently the major contributor of fuel in the form of glucose to the bloodstream. As fasting progresses the adipose tissue becomes the major contributor by supplying free fatty acids and glycerol.

The liver is unique in its ability to contribute fuels to the bloodstream via glycogenolysis, gluconeogenesis, and ketogenesis. These three processes are all inversely related in normal hepatic metabolism: there is a reciprocal relationship between glycogenolysis and gluconeogenesis (both are responsible for total hepatic glucose release), and there is a reciprocal relationship between total hepatic glucose release and ketone-body release. These relationships are illustrated in Figure 2. After an overnight fast the net splanchnic glucose release is 0.86 *mmol/min*/1.73 $m^2$ (Wahren, Felig, Cerasi, and Luft, 1972) in normal adult men. Approximately 80% of this 0.86 *mmol/min*/1.73 $m^2$ can be attributed to hepatic glycogenolysis and the remaining 20% can be attributed to gluconeogenesis derived from lactate, pyruvate, glycerol, and amino acids (Wahren *et al.*, 1972). As fasting is extended glycogenolysis is diminished due to glycogen depletion and gluconeogenesis is augmented. After a three-day fast approximately 80% of the 0.51 *mmol/min*/1.73 $m^2$ net splanchnic glucose release is attributable to gluconeogenesis (Garber *et al.*, 1974). Although there is a reciprocal relationship between the percentage of glucose released either by glycogenolysis or gluconeogenesis, the quantity of glucose supplied by the liver after prolonged fasting by gluconeogenesis is always far less than the quantity transiently supplied by glycogenolysis. However, this deficit in hepatic glucose release is offset by the ability of the liver to make net contributions of the ketone bodies,

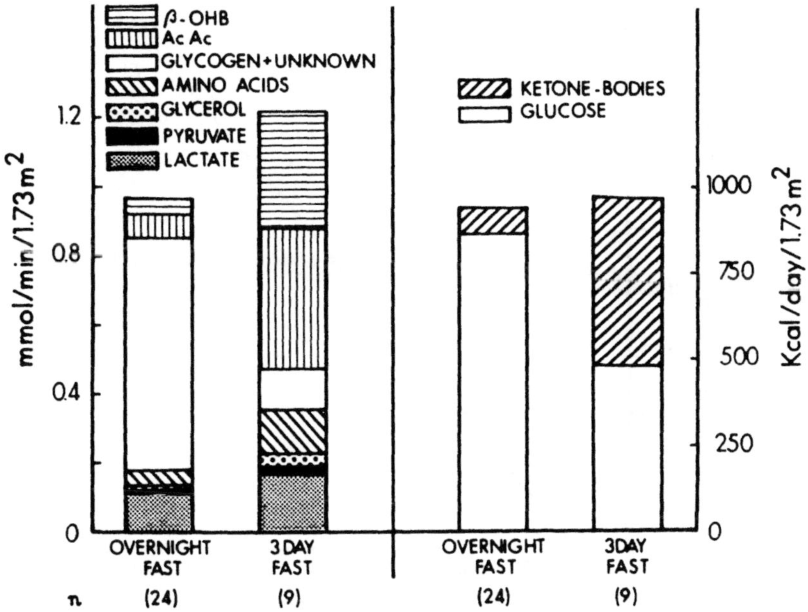


*Figure 2. Net splanchnic glucose and ketone-body release rates, gluconeogenic precursor extraction rates, and caloric equivalents in nondiabetic patients after an overnight and three-day fasts.*

acetoacetate, and β-hydroxybutyrate, to the bloodstream, thereby sparing glucose utilization by peripheral tissues (Owen, Felig, Morgan, Wahren, and Cahill, 1969) and, thus, maintaining glucose homeostasis. The reciprocal relationship between total hepatic glucose release and total hepatic ketone-body release is also shown in Figure 2. After an overnight fast when glucose release is high, ketone-body release is low (0.12 *mmol/min*/1.73 $m^2$). Conversely, after a three-day fast when glucose release is low, ketone-body release is high (0.75 *mmol/min*/1.73 $m^2$)(Garber *et al.*, 1974). Note that the caloric equivalents of fuels (glucose plus ketone bodies) added to the bloodstream by the liver after an overnight and three-day fast are comparable (Figure 2).

## RELATIONSHIP BETWEEN FREE FATTY ACID CONCENTRATION AND UTILIZATION

Besides amino acids, glucose and ketone bodies, long-chain free fatty acids also serve as energy sources to some body tissues devoid or depleted of endogenous fuel during starvation.

Using isotope tracer techniques, free fatty acid turnover and oxidation rates have been studied. A direct relationship between

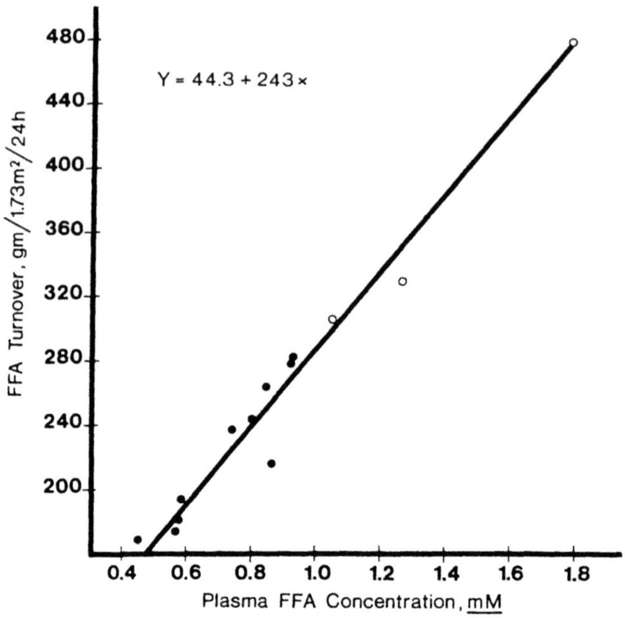


*Figure 3. Relationship between free fatty acid concentration and utilization rate in normal and obese patients. The equation for the line in the figure is derived by the method of least squares. The closed circles are the means of a series of patients and the open circles are single observations.*

free fatty acid turnover and plasma concentration has been observed (Owen and Reichard, 1971)(Figure 3). Further, Issekutz, Paul, Miller, and Bortz (1968) have shown that, regardless of plasma free fatty acid concentration, approximately 30% of the free fatty acid is immediately oxidized. Therefore, some estimate of the contribution of free fatty acids as fuels can be made from knowing plasma free fatty acid concentration.

After an overnight fast the mean concentration of free fatty acids in normal humans is about 0.63 *mmol/l*. From Figure 3 it can be estimated that the free fatty acid turnover rate is about 139 *mg/min* (0.54 *mmol/min*) of which about 30% is immediately oxidized. Assuming a caloric equivalent of 9.3 calories/*mg* of free fatty acids this would provide 0.40 *Kcal/min* or about 50% of the nonprotein calories required by a 150 pound man during the resting state. One to two hours after ingesting breakfast the concentration of plasma free fatty acids decreases to a nadir of about 0.30 *mmol/l*. During this period it can be estimated that free fatty acids provide only

0.15 *Kcal/min* or 19% of the nonprotein calories. However, as more time elapses after food ingestion the concentration of plasma free fatty acids begins to rise slowly as the blood glucose and serum insulin concentrations decrease. The rise in plasma free fatty acids is thought to be due to an increased rate of free fatty release from adipose tissue, consequent to reduced secretion of insulin, and increased secretion of norepinephrine from sympathetic nerve endings in adipose tissue, and perhaps increased secretion of other lipolytic hormones. After three days of starvation plasma free fatty acids plateau with maximum concentrations of about 1.60 *mmol/l* (Owen and Reichard, 1975). At this time it can be calculated that free fatty acids provide 0.90 *Kcal/min* or about 90% of the nonprotein calories. However, it should be recognized that this contribution is derived not only from direct oxidation of free fatty acids *per se* but also from indirect oxidation of ketone bodies after formation from free fatty acids in the liver.

## RELATIONSHIP BETWEEN KETONE-BODY CONCENTRATIONS AND UTILIZATION

After an overnight fast the venous whole blood concentrations of the ketone bodies, acetoacetate (AcAc) and β-hydroxybutyrate (β-OHB), in normal humans are about 0.06 and 0.06 *mmol/l*, respectively. These concentrations decrease only slightly after consuming a mixed meal. However, when obese humans undergo starvation for 4-6 weeks, AcAc increases to about 1.34 *mmol/l* and β-OHB increases to about 5.85 *mmol/l*. The magnitude of these physiologic changes in ketone-body concentrations is the greatest observed of all the fuels available in the bloodstream in the human.

The initial rise in blood ketone bodies during starvation reflects an imbalance between production and removal rates. As starvation is extended, however, blood AcAc and β-OHB plateau at elevated concentrations. After obese humans starve for about 24 days, the rates of production of AcAc and β-OHB are in equilibrium with removal rates, and a new steady state is reached (Owen and Reichard, 1975).

Numerous studies have shown that ketone bodies can be utilized as fuels by a variety of tissues, and that under different conditions AcAc and β-OHB are the principal fuels for muscle (Gammeltoft, 1950; Owen and Reichard, 1971) and kidney (Owen *et al.*, 1969) as well as for the central nervous system (Owen, Morgan, Kemp. Sullivan, Herrera, and Cahill, 1967).

One of the factors that determine ketone-body utilization by tissue is the concentration of these fuels in the circulation. However, factors other than production rate influence the arterial concentration of AcAc and β-OHB. This is demonstrated in Figure 4 which shows results obtained from 45 fasted humans by rapid in-

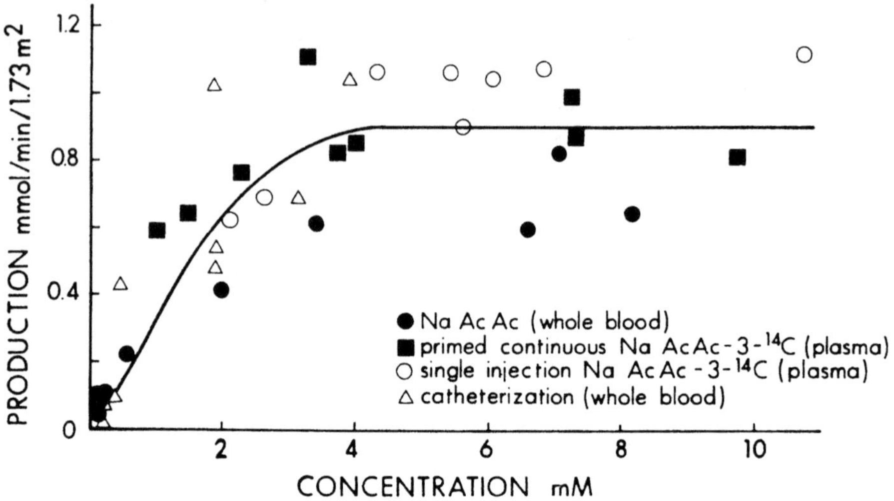


*Figure 4. Relationship between ketone-body concentrations and utilization rates in normal and obese patients undergoing various periods of starvation.*

travenous infusion of large quantities of nonradioactive sodium acetoacetate (Owen, Reichard, Markus, Boden, Mozzoli, and Shuman, 1973), by primed-continuous intravenous infusion (Reichard, Owen, Haff, Paul, and Bortz, 1974) or single-injection of tracer quantities of sodium acetoacetate-3-$^{14}$C (Reichard, unpublished data), and by catheterization techniques (Garber *et al.*, 1974). The curve demonstrates the relationship between total ketone-body concentration in blood and ketone-body production rate as derived by these four diverse techniques. The mean maximum production rate is 0.90 *mmol*/1.73 $m^2$/*min* which extrapolates to about 133 *g*/1.73 $m^2$/24 *h*. A linear relationship between production and concentration is apparent up to 2-4 *mmol*/*l*, a value usually observed after 2-3 days of fasting (Garber *et al.*, 1974). At higher concentrations, such as those observed during more prolonged starvation (Owen and Reichard, 1975), blood concentrations do not reflect production rates. Thus, these combined studies confirm that there is a dissociation between ketone-body production rates and circulating ketone-body concentrations after prolonged starvation when total ketone-body concentration exceeds 4 *mmol*/*l*.

Previous reports have shown that the mean maximum rate of ketone-body oxidation in fasting obese subjects is about 120-130 *g*/24 *h* (Reichard *et al.*, 1974). Thus, after a few days of starvation ketone-body oxidation can contribute about 30%-40% of total caloric requirements. However, it is important to recognize that the contribution of ketone bodies as metabolic fuels to specific tissues changes during starvation.

Ketone bodies serve only transiently as important fuels for skeletal muscle in the human during starvation (Drenick, Alvarez, Tamasi, and Brickman, 1971; Hagenfeldt and Wahren, 1971; Owen and Reichard, 1971). During the first few days of starvation AcAc and β-OHB are the predominant fuels for muscle metabolism. Thereafter, ketone-body consumption by muscle paradoxically decreases despite the marked increase in the arterial concentrations of these substrates with progressive starvation. This is in contradistinction to other tissues such as brain and kidney which show increasing ketone-body uptake with increasing arterial concentrations of these fuels (Gottstein, Müller, Berghoff, Gaertner, and Held, 1971; Hawkins, Williamson, and Krebs, 1971; Little and Spitzer, 1971).

In addition to diminished muscle consumption, another factor responsible for heightened ketonemia observed with starvation is the progressive rise in renal tubular reabsorption of ketone bodies. Both AcAc and β-OHB reabsorption rates increase linearly as the filtered loads of the compounds increase. No tubular maximal transport rates of AcAc and β-OHB have been observed during starvation hyperketonemia (Sapir and Owen, 1975).

Renal conservation of ketone bodies has at least three physiologic effects: (1) since each *mmole* of AcAc or β-OHB excreted in the urine is accompanied by a *mmole* of ammonium, a large quantity of $NH_4$ nitrogen is spared (Owen and Cahill, 1973; Sapir and Owen, 1975); (2) it enhances blood ketone-body concentrations, thus reducing the need for glucose utilization by the central nervous system and, therefore, reduces the requirements for amino acid-derived gluconeogenesis; and (3) hyperketonemia decreases the concentrations of blood alanine and urinary nitrogen excretion during starvation (Sherwin, Hendler, and Felig, 1975), decreases branched-chain amino acids oxidation in muscle (Buse, Biggers, Friderici, and Buse, 1972), and decreases blood alanine and urinary nitrogen excretion after trauma (Blackburn, Flatt, Clowes, O'Donnell, and Hensle, 1973; Smith, Fuller, Wedge, Williamson, and Alberti, 1975). Thus, hyperketonemia resulting from increased hepatic production, diminished muscle uptake, and increased renal conservation, may diminish protein breakdown.

## KETONE-BODY OXIDATION AND ACID-BASE BALANCE

Acetoacetic acid has a *pK* of 3.6 and *beta*-hydroxybutyric acid has a *pK* of 4.4 (Ruderman and Goodman, 1973). Therefore, at physiologic *pH* AcAc and β-OHB anions are over 99% dissociated from their hydrogen cations. In essence, for each *mmole* of AcAc or β-OHB released from the liver, a *mmole* of hydrogen is released from the liver. AcAc and β-OHB production have the potential to affect acid-base balance. However, serious metabolic acidosis does not develop during prolonged starvation (Reidenberg, Haag, Channick, Shuman, and Wilson, 1966; Sapir *et al.*, 1974). In contrast, acidemia is a hallmark of diabetic ketoacidosis. Yet the hepatic production rate of AcAc and β-OHB are practically the same in humans after three or more days of starvation and during diabetic ketoacidosis (Owen, Block, Patel, Boden, McDonough, Kreulen, Shuman, and Reichard, 1977). The difference in acidemia observed between nondiabetic fasting humans and diabetic ketoacidotic patients may reside primarily in different rates of ketone-body oxidation.

Athough there is a dissociation between net splanchnic ketone-body production rates and peripheral blood ketone-body concentrations after the blood concentration exceeds 2-4 *mmoℓ/ℓ*, during starvation production rates of AcAc and β-OHB eventually equilibrate with removal rates and a steady state is achieved (Owen and Reichard, 1975). In uncontrolled diabetic ketoacidosis, ketone-body production and removal rates do not equilibrate, and death results from progressive ketoacidosis. The primary route of ketone-body removal during starvation is tissue oxidation (Owen and Reichard, 1975). Several investigators have reported impaired ketone-body oxidation rates in diabetic animals during ketoacidosis (Balasse and Havel, 1971; Beatty, Peterson, Bocek, and West, 1959; Ruderman and Goodman, 1974), and we have made a similar observation in one diabetic patient during ketoacidosis (Owen and Reichard, 1975). Table 3 compares total ketone-body production and oxidation rates in obese fasted individuals with one diabetic ketoacidotic patient.

The mean plasma ketone-body concentration in the diabetic patient was greater than that in obese individuals. The measured

*Table 3. KETONE-BODY PRODUCTION AND OXIDATION RATES IN OBESE FASTED SUBJECTS AND A DIABETIC KETOACIDOTIC PATIENT USING ISOTOPE TRACER TECHNIQUES*

| | Concentration | Production | | Oxidation | |
|---|---|---|---|---|---|
| | *mmoℓ/ℓ* | *mmoℓ/ℓ* | *mmoℓ*/1.73 $m^2$/*min* | *mmoℓ/min* | *mmoℓ*/1.73 $m^2$/*min* |
| Obese fasting (3)* | 6.5 | 1.09 | 0.96 | 0.95 | 0.83 |
| Diabetic ketoacidosis (1) | 12.44 | 1.08 | 1.39 | 0.16 | 0.21 |

**The number in parenthesis is the number of subjects (or patients) studied. The obese subjects were fasted for 24 days, and the values given are the means. The diabetic subject was in severe ketoacidosis with a serum bicarbonate concentration of 10 mmoℓ/ℓ.*

production rates in *mmol/min* in obese subjects and the diabetic patient are comparable. When production rates were corrected to 1.73 $m^2$ body surface area, the diabetic patient's rate exceeded the obese subjects' rate. In contrast, the measured or corrected oxidation rates in the diabetic ketoacidotic patient were less than one-third those observed in obese humans. An overview of Table 3 shows that measured production rates were comparable in the obese subjects and in the ketoacidotic patient while measured or corrected oxidation rate was greater in fasted obese individuals. Thus, it appears that diabetic patients with hyperketonemia have decreased rates of ketone-body oxidation while the rates of ketone-body production and oxidation were approximately equal in the obese subjects.

It appears: (1) that nondiabetic obese humans produce and oxidize approximately equal quantities of ketone bodies during starvation; (2) that the diabetic patient in ketoacidosis produces approximately the same quantities of ketone bodies as the obese starving humans produce; and (3) that the diabetic patient in ketoacidosis oxidizes less ketone bodies than did the obese starving humans. This may explain the differences observed in the degree of hyperketonemia seen in these two patient populations. However, the question to be answered is, "How does the difference in rates of ketone-body oxidation explain the difference in the degrees of acidemia in these two groups?" Perhaps the answer is simple. If an acid, *e.g.*, acetoacetic acid or *beta*-hydroxybutyric acid, is produced and then oxidized, acid-base balance is not disturbed. This implies that the cation (hydrogen) and anion (acetoacetate, *beta*-hydroxybutyrate) are cleared jointly through catabolism. This is shown schmatically in Figure 5 which suggests: (1) that acetoacetic acid (or *beta*-hydroxybutyric acid) is dissociated

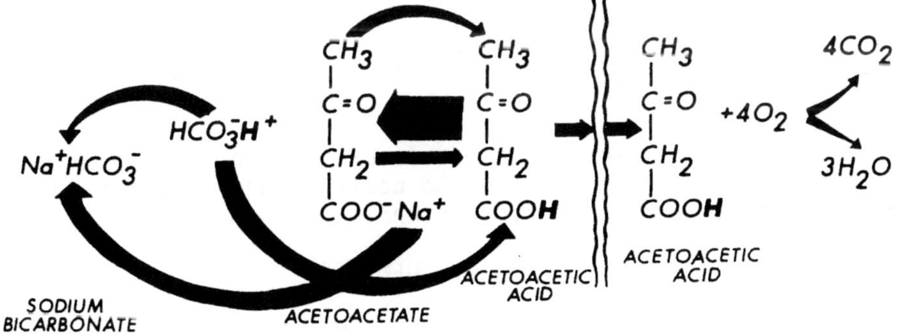


*Figure 5. Schematic speculation on ketone-body catabolism. Note that hydrogen ion is incorporated during transfer and catabolism.*

from hydrogen in the extracellular spaces; (2) that acetoacetic acid crosses the cellular membrane in the undissociated form (then redissociates in the cytoplasm); and (3) that during the catabolism of acetoacetate to $CO_2$ and $H_2O$, acetoacetate and hydrogen are consumed simultaneously. This does not imply that the enzymes catabolize acetoacetic acid, but it does suggest that one *mole* of hydrogen is removed from the mitochondrial matrix when one *mole* of anion (acetoacetate or *beta*-hydroxybutyrate) is catabolized to $CO_2$ and $H_2O$.

SUMMARY

During feasting on a balanced carbohydrate, fat, and protein meal resting metabolic rate, body temperature and respiratory quotient all increase. The dietary components are utilized to replenish and augment glycogen and fat stores in the body. Excessive carbohydrate is also converted to lipid in the liver and stored along with the excessive lipids of dietary origin as triglycerides in adipose tissue, the major fuel storage depot. Amino acids in excess of those needed for protein synthesis are preferentially catabolized over glucose and fat for energy production. This occurs because there are no significant storage sites for amino acids or proteins, and the accumulation of nitrogenous compounds is ill tolerated.

During fasting, adipose tissue, muscle, liver, and kidneys work in concert to supply, to convert, and to conserve fuels for the body. During the brief postabsorptive period, blood fuel homeostasis is maintained primarily by hepatic glycogenolysis and adipose tissue lipolysis. As fasting progresses, muscle proteolysis supplies glycogenic amino acids for heightened hepatic gluconeogenesis for a short period of time. After about three days of starvation, the metabolic profile is set to conserve protein and to supply greater quantities of alternate fuels. In particular, free fatty acids and ketone bodies are utilized to maintain energy needs. The ability of the kidney to conserve ketone bodies prevents the loss of large quantities of these valuable fuels in the urine.

This delicate interplay among liver, muscle, kidney, and adipose tissue maintains blood fuel homeostasis and allows humans to survive caloric deprivation for extended periods.

ACKNOWLEDGEMENTS

The authors gratefully acknowledge Ms. M. Mozzoli and Ms. P. Popkave for their technical assistance and Ms. R. Matza for typing this manuscript.

This work was supported, in part, by U.S. Public Health Service grant 5 M01 RR 349-11; NIH General Clinical Research Centers Branch, NIH Arthritis Metabolic and Digestive Diseases grants AM 16102, AM 19397, and AM 13527; and NIH Neurological and Communicative Disorders and Stroke grant NS 11088.

## REFERENCES

Adibi, S.A. (1976). Metabolism of branched-chain amino acids in altered nutrition. *Metabolism* 25:1287-1302.

Ahmed, M., Gannon, M.C., and Nuttall, F.Q. (1976). Postprandial plasma glucose, insulin, glucagon and triglyceride responses to a standard diet in normal subjects. *Diabetologia* 12:61-67.

Balasse, E.O. and Havel, R.J. (1971). Evidence for an effect of insulin on the peripheral utilization of ketone bodies in dogs. *J. Clin. Invest.* 50:801-813.

Barter, P.J., Nestel, P.J., and Carroll, K.F. (1972). Precursors of plasma triglyceride fatty acid in humans. Effects of glucose consumption, clofibrate administration, and alcoholic fatty liver. *Metabolism* 21:117-124.

Beatty, C.H., Peterson, R.D., Bocek, R.M., and West, E.S. (1959). Acetoacetate and glucose uptake by diaphragm and skeletal muscle from control and diabetic rats. *J. Biol. Chem.* 234: 11-15.

Benedict, F.G. (1915). A study of prolonged fasting. Carnegie Institute of Washington, Washington, DC, publication 203.

Blackburn, G.L., Flatt, J.P., Clowes, G.H.A. Jr., O'Donell, T.F., and Hensle, T.E. (1973). Protein sparing therapy during periods of starvation with sepsis or trauma. *Annals of Surgery* 177:588-594.

Boberg, J., Carlson, L.A., Freyschuss, U., Lassers, B.W., and Wahlquist, M.L. (1972). Splanchnic secretion rates of plasma triglycerides and total and splanchnic turnover of plasma free fatty acids in men with normo- and hypertriglyceridaemia. *Eur. J. Clin. Invest.* 2:454-466.

Buse, M.G., Biggers, J.F., Friderici, K.H., and Buse, J.F. (1972). Oxidation of branched chain amino acids by isolated hearts and diaphragms of the rat. *J. Biol. Chem.* 247:8085-8096.

Cahill, G.F. Jr., Herrera, M.G., Morgan, A.P., Soeldner, J.S., Steinke, J., Levy, P.L., Reichard, G.A. Jr., and Kipnis, D.M. (1966). Hormone-fuel interrelationships during fasting. *J. Clin. Invest.* 45:1751-1769.

Drenick, E.J., Alvarez, L.C., Tamasi, G.C., and Brickman, A.S. (1972). Resistance to symptomatic insulin reactions after fasting. *J. Clin. Invest.* 51:2757-2762.

Dubois, E.F. (1921). The basal metabolism in fever. *JAMA* 77:352-357.

Felig, P. (1973). The glucose-alanine cycle. *Metabolism* 22:179-207

Felig, P., Owen, O.E., Wahren, J., and Cahill, G.F. Jr. (1969). Amino acid metabolism during prolonged starvation. *J. Clin. Invest.* 48:584-594.

Felig, P., Wahren, J., and Hendler, R. (1975). Influence of oral glucose ingestion on splanchnic glucose and gluconeogenic substrate metabolism in man. *Diabetes* 24:468-475.

Gammeltoft, A. (1950). The significance of ketone bodies in fat metabolism. I. Concentration of ketone bodies in the arterial and venous blood in human subjects during starvation. *Acta Physiol. Scand.* 19:270-287.

Garber, A.J., Menzel, P.H., Boden, G., and Owen, O.E. (1974). Hepatic ketogenesis and gluconeogenesis in humans. *J. Clin. Invest.* 54:981-989.

Goldman, R.F., Haisman, M.F., Bynum, G., Horton, E.S., and Sims, E.A.H. (1973). Experimental obesity in man: Metabolic rate in relation to dietary intake. pp. 165-186. In G.A. Bray (Ed.) *Obesity in Perspective.* Fogarty International Center Series on Preventive Medicine, DHEW Publication No. (NIH) 75-708.

Gottstein, U., Müller, W., Berghoff, W., Gartner, H., and Held, K. (1971). Zur Utilisation von nicht-veresterten Fettsauren und Ketonkorpern im Gehirn des Menschen. *Klin. Wchr.* 49:406-411.

Hagenfeldt, L. and Wahren, J. (1971). Human forearm muscle metabolism during exercise. VI. Substrate utilization in prolonged fasting. *Scand. J. Clin. Lab. Invest.* 27:299-306.

Havel, R.J. (1972). Caloric homeostasis and disorders of fuel transport. *N. Eng. J. Med.* 287:1186-1192.

Havel, R.J., Kane, J.P., Balasse, E.O., Segel, N., and Basso, L.V. (1970). Splanchnic metabolism of free fatty acids and production of triglycerides of very low density lipoproteins in normotriglyceridemic and hypertriglyceridemic humans. *J. Clin. Invest.* 49:2017-2035.

Hawkins, R.A., Williamson, D.H., and Krebs, H.A. (1971). Ketone-body utilization by adult and suckling rat brain *in vivo.* *Biochem. J.* 122:13-18.

Issekutz, B. Jr., Paul, P., Miller, H.I., and Bortz, W. (1968). Oxidation of plasma free fatty acids in lean and obese humans. *Metabolism* 17:62-73.

Keys, A., Anderson, J.T., and Brozek, J. (1955). Weight gain from simple overeating. *Metabolism* 4:427-432.

Krebs, H.A. (1972). Some aspects of the regulation of fuel supply in omnivorous animals. *Adv. Enzyme Regul.* 10:397-420.

Little, J.R. and Spitzer, J.J. (1971). Uptake of ketone bodies by dog kidney *in vivo.* *Am. J. Physiol.* 221:679-683.

Marliss, E.B., Aoki, T.T., Pozefsky, T., Most, A.S., and Cahill, G.F. (1971). Muscle and splanchnic glutamine and glutamate metabolism in postabsorptive and starved man. *J. Clin. Invest.* 50:814-817.

Müller, W.A., Faloona, G.R., Aguilar-Parada, E., and Unger, R.H. (1970). Abnormal alpha-cell function in diabetes. Response to carbohydrate and protein ingestion. *N. Eng. J. Med.* 283: 109-115.

Nilsson, L.H. (1973). Liver glycogen content in man in the post-absorptive state. *Scand. J. Clin. Lab. Invest.* 32:317-323.

Nilsson, L.H. and Hultman, E. (1974). Liver and muscle glycogen in man after glucose and fructose infusion. *Scand. J. Clin. Invest.* 33:5-10.

Owen, O.E., Block, B.S.B., Patel, M., Boden, G., McDonough, M., Kreulen, T., Shuman, C.R., and Reichard, G.A. Jr. (1977). Human splanchnic metabolism during diabetic ketoacidosis. *Metabolism* 26:381-398.

Owen, O.E. and Cahill, G.F. Jr. (1973). Metabolic effects of exogenous glucocorticoids in fasted man. *J. Clin. Invest.* 52:2596-2605.

Owen, O.E., Felig, P., Morgan, A.P., Wahren, J., and Cahill, G.F. Jr. (1969). Liver and kidney metabolism during prolonged starvation. *J. Clin. Invest.* 48:574-583.

Owen, O.E., Morgan, A.P., Kemp, H.G., Sullivan, J.M., Herrera, M.G., and Cahill, G.F. Jr. (1967). Brain metabolism during fasting. *J. Clin. Invest.* 46:1589-1595.

Owen, O.E., Patel, M.S., Block, B.S.B., Kreulen, T.H., Reichle, F.A., and Mozzoli, M.A. (1976). Gluconeogenesis in normal, cirrhotic and diabetic humans. In R.W. Hanson and M.A. Mehlman (Eds.) *Gluconeogenesis: Regulation in Mammalian Species.* John Wiley, New York.

Owen, O.E. and Reichard, G.A. Jr. (1975). Ketone body metabolism in normal, obese and diabetic subjects. *Israel J. Med. Sci.* 11:560-570.

Owen, O.E. and Reichard, G.A. Jr. (1971). Human forearm metabolism during prolonged starvation. *J. Clin. Invest.* 50:1536-1545.

Owen, O.E., Reichard, G.A. Jr., Markus, H., Boden, G., Mozzoli, M.A., and Shuman, C.R. (1973). Rapid intravenous sodium acetoacetate infusion in man. Metabolic and kinetic responses. *J. Clin. Invest.* 52:2606-2615.

Passmore, R. and Swindells, Y.E. (1963). Observations on the respiratory quotients and weight gains of man after eating large quantities of carbohydrate. *Brit. J. Nutr.* 17:331-339.

Patel, M.S., Owen, O.E., Goldman, L.I., and Hanson, R.W. (1975). Fatty acid synthesis by human adipose tissue. *Metabolism* 24: 161-173.

Reichard, G.A. Jr., Owen, O.E., Haff, A.C., Paul, P., and Bortz, W.M. (1974). Ketone-body production and oxidation in fasting. *J. Clin. Invest.* 53:508-515.

Reidenberg, M.M., Haag, B.L., Channick, B.J., Shuman, C.R., and Wilson, T.G.G. (1966). The response of bone to metabolic acidosis in man. *Metabolism* 15:236-241.

Ruderman, N.B. and Goodman, M.N. (1974). Inhibition of muscle acetoacetate utilization during diabetic ketoacidosis. *Am. J. Physiol.* 226:136-143.
Ruderman, N.B. and Goodman, M.N. (1973). Regulation of ketone body metabolism in skeletal muscle. *Am. J. Physiol.* 224:1391-1397.
Sapir, D.G. and Owen, O.E. (1975). Renal conservation of ketone bodies during starvation. *Metabolism* 24:23-33.
Sapir, D.G., Owen, O.E., Pozefsky, T., and Walser, M. (1974). Nitrogen sparing induced by a mixture of essential amino acids given chiefly as their ketoanalogues during prolonged starvation in obese subjects. *J. Clin. Invest.* 54:974-980.
Sherlock, S. (1975). Anatomy of the liver. p. 1. In S. Sherlock (Ed.) *Diseases of the Liver and Biliary System*, 5th ed. Blackwell Scientific Publications, Oxford.
Sherwin, R.S., Hendler, R.G., and Felig, P. (1975). Effect of ketone infusions on amino acid and nitrogen metabolism in man. *J. Clin. Invest.* 55:1382-1390.
Smith, R., Fuller, D.J., Wedge, J.H., Williamson, D.H., and Alberti, K.G.G.M. (1975). Initial effect of injury on ketone bodies and other blood metabolites. *Lancet* I:1-3.
Wahren, J., Felig, P., Cerasi, E., and Luft, R. (1972). Splanchnic and peripheral glucose and amino acid metabolism in diabetes mellitus. *J. Clin. Invest.* 51:1870-1878.
Wahren, J., Felig, P., and Hagenfeldt, L. (1976). Effect of protein ingestion on splanchnic and leg metabolism in normal man and in patients with diabetes mellitus. *J. Clin. Invest.* 57:987-999.
Wilmore, D.W. and Dudrick, S.J. (1968). Growth and development of an infant receiving all nutrients exclusively by vein. *JAMA* 203:140-144.

## ADDENDUM

During the discussion, Dr. E.A.H. Sims was invited to comment on his research on overfeeding. The findings are reviewed in:

Sims, E.A.H. (1976). Experimental obesity, dietary-induced thermogenesis, and their clinical implications. *Clinics Endocrinol. Metab.* 5:377-395.

SUBJECT INDEX

GPSR Compliance

*The European Union's (EU) General Product Safety Regulation (GPSR) is a set of rules that requires consumer products to be safe and our obligations to ensure this.*

*If you have any concerns about our products, you can contact us on ProductSafety@springernature.com*

In case Publisher is established outside the EU, the EU authorized representative is:

Springer Nature Customer Service Center GmbH
Europaplatz 3
69115 Heidelberg, Germany

**Batch number: 10620881**

Printed by Printforce, the Netherlands